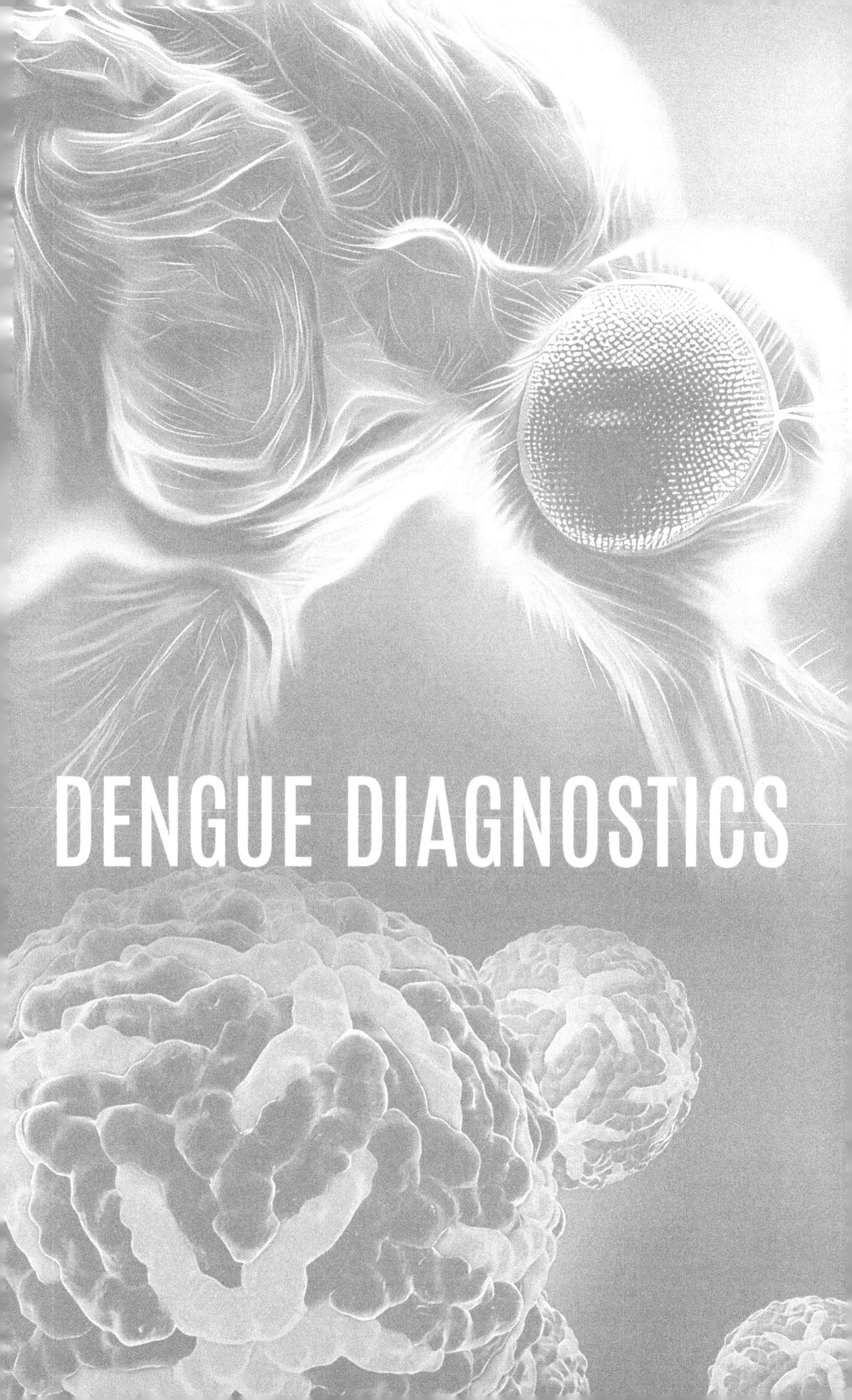

DENGUE DIAGNOSTICS

DENGUE DIAGNOSTICS

The Right Test at the Right Time for the Right Group

edited by

Shamala Devi Sekaran

JENNY STANFORD
PUBLISHING

Published by

Jenny Stanford Publishing Pte. Ltd.
101 Thomson Road
#06-01, United Square
Singapore 307591

Email: editorial@jennystanford.com
Web: www.jennystanford.com

British Library Cataloguing-in-Publication Data
A catalogue record for this book is available from the British Library.

Dengue Diagnostics: The Right Test at the Right Time for the Right Group

ISBN 978-981-4968-97-3 (Hardcover)
ISBN 978-1-032-66956-4 (eBook)

Contents

Preface

Dengue is a mosquito-borne viral infection that affects millions of people worldwide every year. It can cause a range of symptoms, from mild fever and rash to severe hemorrhagic fever and shock. Dengue is a major public health problem, especially in tropical and subtropical regions, where it poses a significant burden on health systems and economies.

The diagnosis of dengue is crucial for the timely management and treatment of patients, as well as for the prevention and control of outbreaks. The diagnosis of dengue is challenging due to the complexity of the disease, the diversity of the virus, and the limitations of the available diagnostic methods. The clinical manifestations are often nonspecific and overlap with other febrile illnesses. Moreover, the laboratory confirmation of dengue requires the detection of either the virus, its antigens, or its antibodies, which have different sensitivities and specificities at different stages of the infection. Therefore, there is a need for accurate, rapid, and affordable diagnostic tests that can be used in various settings and scenarios.

This book is based on my extensive experience and research in the field of dengue diagnostics as well as that of the experts and collaborators who have contributed to the chapters. It provides a comprehensive overview of the current state of the art in dengue diagnostics, covering the principles, techniques, applications, and challenges of various methods. It also discusses future directions and perspectives for the development and improvement of dengue diagnostics.

The book is intended for researchers, clinicians, public health professionals, students, and anyone interested in learning more about dengue diagnostics, and I hope it will be a valuable resource and reference for them and will inspire further research and innovation in this important area of global health.

Shamala Devi Sekaran

Summer 2023

Chapter 1

Introduction to Dengue and Issues with Current Diagnostics

Shamala Devi Sekaran,[a] Rishya Manikam,[b] and Gaythri Thergarajan[a]
[a]*Faculty of Applied Sciences, UCSI University Kuala Lumpur Campus, Malaysia*
[b]*Emergency and Trauma Centre, University Malaya Medical Centre, Kuala Lumpur, Malaysia*
shamala@ucsiuniversity.edu.my

Historically dengue is said to have been derived from the Swahili phrase "Ki-denga pepo", which basically translates to mean cramp-like seizures caused by an evil spirit. The Swahili word may possibly originate from the Spanish word "dengue", which actually describes the gait of a person suffering from intense bone pain (Hotta, 1952; Skae, 1902). Today, however, we know dengue as a viral infection caused by the dengue virus belonging to the family Flaviviridae. The first record of this illness is recorded in the Chinese medical encyclopedia during the Jin Dynasty (265–420) and the illness was referred to as a water poison associated with flying insects. The

Dengue Diagnostics: The Right Test at the Right Time for the Right Group
Edited by Shamala Devi Sekaran

ISBN 978-981-4968-97-3 (Hardcover), 978-1-032-66956-4 (eBook)
www.jennystanford.com

first report of dengue disease was reported in Philadelphia in 1780. First, recognized epidemics were noted to occur simultaneously in Asia, Africa, and North America in the 1780s. Upon identification, the disease was named with the first confirmed case report in the same year by Benjamin Rush who was the one who coined the name "breakbone fever" because of the symptoms of myalgia and arthralgia. Japanese scientists were the first to discover the DENV-1 in 1943. Subsequently, in 1944, DENV-2 was discovered by Albert Sabin. In Malaysia, Skae reported the first known published outbreak of dengue in 1902, while severe dengue was first observed in Georgetown Penang in 1962 (Skae, 1902).

Dengue is a mosquito-borne viral infection caused by the dengue virus, the genus of which contains four serotypes. While this virus causes mild infections with flu-like illness, it can occasionally develop into a potentially lethal complication called severe dengue. Found mainly in tropical and sub-tropical climates and mostly in urban and semi-urban communities, it has, however, spread over the temperate zones as a result of climate change. There is currently no specific treatment but early detection and access to proper medical care can lower the fatality rate to below 1%. The global incidence as seen in Table 1.1 indicates the populations at risk and this has been on the rise, and it is currently estimated that about half of the world's population is now at risk (Figure 1.1). Currently, prevention and control depend on stringent effective vector control measures with the help of community participation.

1.1 Current Epidemiology

In 2022, more than 2357301 cases and 1731 deaths were reported. Most of the cases recorded were from Brazil, Vietnam, Peru, the Philippines, and Indonesia; and most deaths were reported from Brazil, Indonesia, the Philippines, Peru, and Timor-Leste. All four dengue serotypes were detected. In Europe, no autochthonous case of dengue was reported in 2022 (European Centre for Disease Prevention and Control [ECDC] 2022).

Table 1.1 Dengue cases around the globe, as of July 2022 (Extracted from https://www.ecdc.europa.eu/en/dengue-monthly)

Year	Date	Country	Number of cases	Number of deaths
2022	9 July	Afghanistan	22	0
2022	26 July	Bangladesh	6673	8
2022	14 July	Cambodia	3322	9
2022	31 May	China	5	0
2022	9 July	Indonesia	52313	448
2022	14 July	Laos	6393	10
2022	14 July	Malaysia	26420	19
2022	31 May	Maldives	344	0
2022	24 July	Nepal	232	0
2022	7 April	Oman	76	0
2022	4 July	Pakistan	875	0
2022	25 June	Philippines	64797	274
2022	25 July	Singapore	21350	0
2022	25 July	Sri Lanka	32404	0
2022	19 July	Thailand	8660	0
2022	27 May	Timor-Leste	4985	56
2022	14 July	Vietnam	103433	37
2022	28 April	Kenya	33	0
2022	14 July	Australia	66	0
2022	28 May	Cook Islands	3	0
2022	4 June	Micronesia	16	0
2022	14 July	Palau	22	0
2022	7 May	Solomon Islands	34	0
2022	16 June	Vanuatu	39	0
2022	16 June	Wallis & Futuna	21	0
2022	27 July	Brazil	1827617	737
2022	27 July	Peru	56021	65
2022	27 July	Colombia	32327	NA
2022	27 July	Nicaragua	22887	NA
2022	27 July	Ecuador	12279	NA

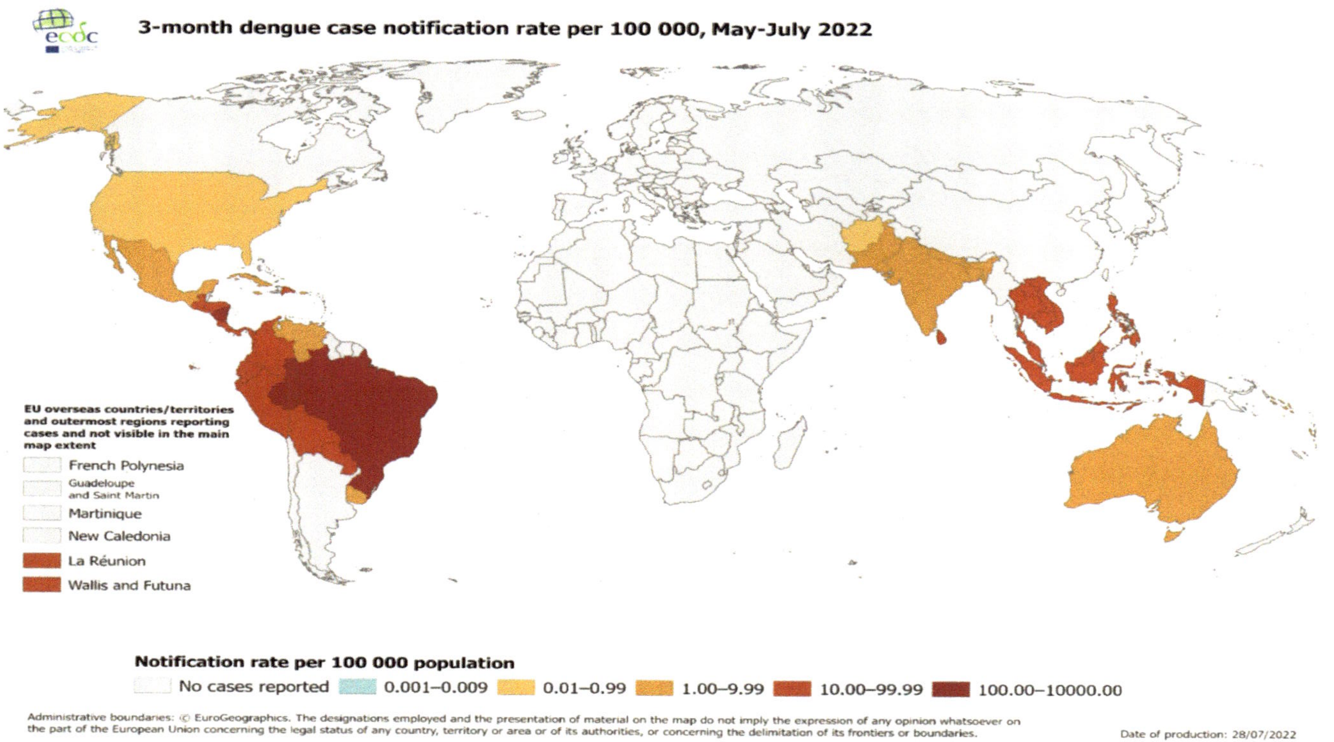

Figure 1.1 Geographical distribution of dengue cases reported worldwide, May to July, 2022.

1.2 Diagnosis

Dengue can be diagnosed clinically; however, laboratory tests are required to confirm the infection. Definitive diagnosis is important not only for the clinical management of patients, but also for interventions during outbreaks, epidemiological surveillance, and for vaccine development and monitoring. The development of assays is ongoing and newer assays using more advanced technologies have been developed, though not fully validated. The main hurdle lies in the incompletely understood pathogenesis of dengue and that multiple sequential infections occur in dengue-endemic areas, notwithstanding the issues with primary and secondary infection status that further complicate the diagnosis. Therefore, for a diagnostic assay to be useful and effective, it is essential for users to have some degree of confidence in the test in order to improve disease management, especially in the acute early stage and for detecting signs of severity. An ideal dengue diagnostic test would be one that is simple, rapid, with high sensitivity and specificity, preferably able to differentiate between primary and secondary infections, serotype the virus, and most important of all affordable. The optimal time frame for diagnosis would be from the onset to 10 days post-infection. In reality, more than half only consult the physician when in dire situations, with many people in third-world countries resorting to traditional healing. It is also important to note that 2% do not seroconvert and a large number are asymptomatic.

Diagnostic tools currently are mainly serologically based, nucleic acid-based, or antigen detection. A good understanding of the clinical conditions of dengue patients is essential for the appropriate usage of these tests. The highest confidence lies in the isolation of the virus, thus fulfilling Koch's postulate, detection by direct immunofluorescence (IF), and genome amplification via PCR (Medina et al., 2012; Prescott, Feldmann, & Safronetz, 2017; Salles et al., 2018; World Health Organization, 2009), or detection of viral antigens like NS1 using either ELISAs or rapid lateral flow tests (Chong et al., 2020; Guzman, Jaenisch et al., 2010; Hunsperger et al., 2009; Kumarasamy, Chua et al., 2007; Kumarasamy, Wahab et al., 2007; Peeling et al., 2010; S. D. Sekaran, Lan, & Subramaniam, 2018; Shamala Devi, 2008; Tricou et al., 2010; Wang & Sekaran, 2010b, 2010a). However, more often than not we tend to use tests out of convenience and hence look for antibodies, and for these, diagnostically useful results are best obtained using paired samples. Currently, virus isolation, PCR, and direct IF are carried out only in reference and research laboratories. NS1, however, is more user-friendly, and apart from ELISAs, rapid flow tests have been developed and validated. These, however, are useful in the first 5 days of symptoms when viremia still exists. Serology can also be utilized by demonstrating a seroconversion from negative to positive IgM antibody to dengue, or a four-fold or greater increase in IgG antibody titers in paired (acute and convalescent) serum specimens. Patients who are IgM positive and PCR negative are usually classified as recent probable dengue infection as IgM antibodies to dengue remain elevated for 90 days after the illness and could have been from an infection that occurred 2–3 months ago. In addition, it is important to note that cross-reactivity with other flaviviruses including West Nile virus (WNV), St. Louis encephalitis virus (SLE), Japanese encephalitis virus (JEV), Zika virus, and yellow fever virus (YFV) do occur and hence a review of the patient's past medical history, recent travel history, and vaccination record (especially yellow fever vaccination) are needed to determine the likelihood that the current acute febrile illness is actually due to an infection with dengue virus. In many instances, a paired sample is best utilized to make a diagnosis, especially for those who present late after the onset of illness (> 5 days), as virus and viral antigens become undetectable, and in these instances, the tests need to be carried out on all viruses suspected. The current infection will be the only pair demonstrating a four-fold rise (Schilling, Ludolfs, Van An, & Schmitz, 2004; Shamala Devi Sekaran & Soe, 2017).

1.3 Immune Responses to Dengue

Dengue virus induces the production of all classes of antibodies primarily targeting the virus envelope proteins. The level of antibodies primarily depends on whether the individual has primary or secondary dengue infection (S. D. Sekaran, Ew, Subramaniam, & Kanthesh, 2009; Shamala Devi Sekaran & Artsob, 2007; Uno & Ross, 2018). Figure 1.2 shows the timelines of the disease and Figure 1.3 displays the production of the various aspects of the immune response.

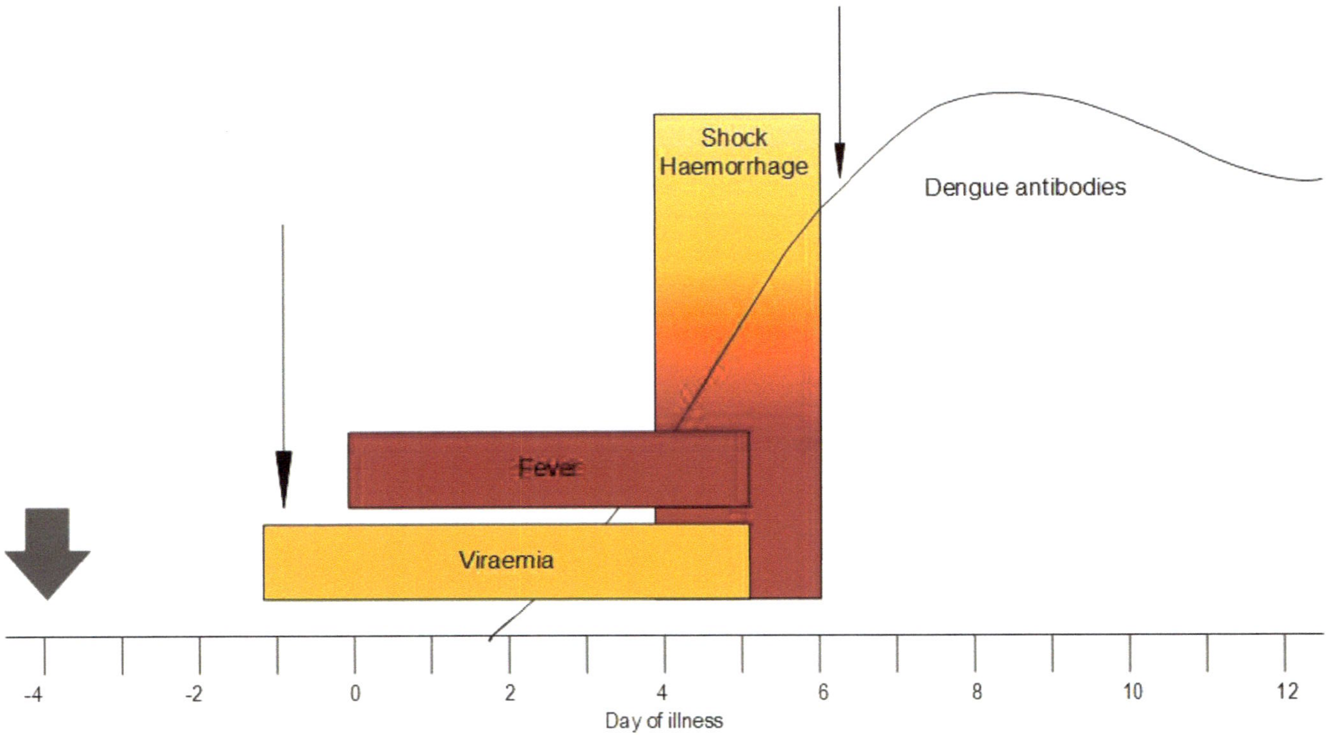

Figure 1.2 Timeline of the various components of the dengue infection.

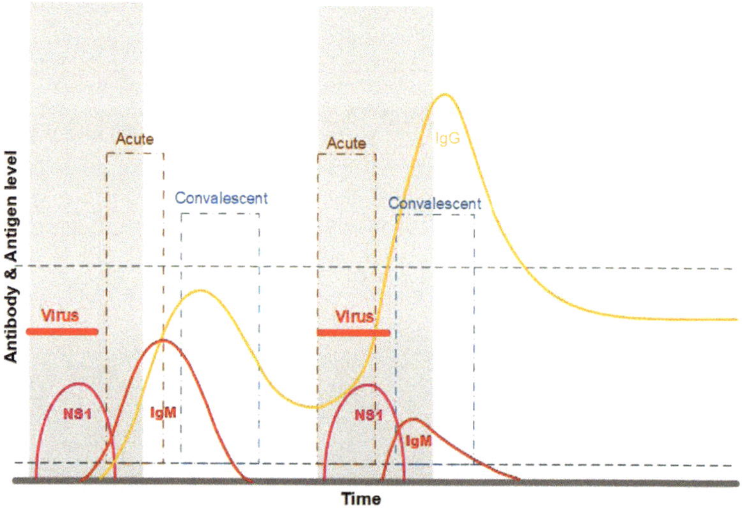

Figure 1.3 Diagnostic timeline of primary and secondary immune responses.

In recent years, the development of rapid assays has allowed patient specimens to be tested in point-of-care situations. Many manufacturers of PCTs claim the tests are able to detect and differentiate between primary and secondary infections with dengue (Changal et al., 2016; Lovera et al., 2019). However, the diagnostic accuracy of such assays has not been reliably established because of the multiplicity of methodologies used in the evaluation. Newer bio-diagnostic devices that can be quantitative and/or qualitative, are the prototypes of the future rapid diagnostic test kits that will be commercialized if they have desirable traits such as the ability to be portable, automated, and easily disposed of. A survey of the literature shows that most biosensors that are being developed for dengue use piezoelectric, optical, and electrochemical methods (Eivazzadeh-Keihan et al., 2019). More recently, nanoparticle beads, mass spectrometry, as well as micro-sequencing have been utilized and appeared promising. Generally, biosensor kits need to be validated to be used as a rapid test for dengue.

1.4 Markers and Problems

Diagnosis begins with a clinical suspicion but has limited usefulness for early diagnosis. Clinical symptoms have high sensitivities but have poor specificities and hence a confirmatory diagnosis is essential. The current markers used are the non-structural antigen-1 (NS1), IgM, and IgG antibodies, while virus isolation and nucleic acid amplification are only utilized in laboratories tracking infections or public health laboratories. The titers of the IgM and IgG antibodies depend on whether the infection is a primary or secondary infection. During primary (first) dengue infection, IgM levels are very high, but during secondary infection, IgM levels are lower. The levels of IgG actually increase during secondary infection. Clinicians, therefore, need to measure IgM and IgG quantitatively in order to determine whether a patient has primary or secondary dengue infection. The main issue with confirmatory assays is the validation of each assay. Hence, early diagnosis remains a challenge, more so markers to use, and the time frame of detection. Using a combination of clinical markers and laboratory-confirmed assays may serve as predictive markers at different stages of the disease.

References

Changal, K. H., Raina, A. H., Raina, A., Raina, M., Bashir, R., Latief, M., ... Changal, Q. H. (2016). Differentiating secondary from primary dengue using IgG to IgM ratio in early dengue: An observational hospital based clinico-serological study from North India. *BMC Infectious Diseases, 16*(1), 715. https://doi.org/10.1186/s12879-016-2053-6

Chong, Z. L., Sekaran, S. D., Soe, H. J., Peramalah, D., Rampal, S., & Ng, C.-W. (2020). Diagnostic accuracy and utility of three dengue diagnostic tests for the diagnosis of acute dengue infection in Malaysia. *BMC Infectious Diseases, 20*(1), 210. https://doi.org/10.1186/s12879-020-4911-5

Eivazzadeh-Keihan, R., Pashazadeh-Panahi, P., Mahmoudi, T., Chenab, K. K., Baradaran, B., Hashemzaei, M., ... Maleki, A. (2019). Dengue virus: A review on advances in detection and trends – from conventional methods to novel biosensors. *Microchimica Acta, 186*(6), 329. https://doi.org/10.1007/s00604-019-3420-y

European Centre for Disease Prevention and Control (ECDC) 2022. (2022). European Centre for Disease Prevention and Control: Dengue Worldwide Overview. Retrieved from https://www.ecdc.europa.eu/en/dengue-monthly

Guzman, M. G., Halstead, S. B., Artsob, H., Buchy, P., Farrar, J., Gubler, D. J., ... Peeling, R. W. (2010). Dengue: A continuing global threat. *Nature*

Reviews Microbiology, 8(S12), S7–S16. https://doi.org/10.1038/nrmicro2460

Hotta, S. (1952). Experimental studies on dengue: Isolation, identification and modification of the virus. *Journal of Infectious Diseases, 90*(1), 1–9. https://doi.org/10.1093/infdis/90.1.1

Hunsperger, E. A., Yoksan, S., Buchy, P., Nguyen, V. C., Sekaran, S. D., Enria, D. A., ... Peeling, R. W. (2009). Evaluation of commercially available anti-dengue virus immunoglobulin M tests. *Emerging Infectious Diseases, 15*(3), 436–440. https://doi.org/10.3201/eid1503.080923

Kumarasamy, V., Chua, S. K., Hassan, Z., Wahab, A. H. A., Chem, Y. K., Mohamad, M., & Chua, K. B. (2007). Evaluating the sensitivity of a commercial dengue NS1 antigen-capture ELISA for early diagnosis of acute dengue virus infection. *Singapore Medical Journal, 48*(7), 669–673.

Lovera, D., Martínez-Cuellar, C., Galeano, F., Amarilla, S., Vazquez, C., & Arbo, A. (2019). Clinical manifestations of primary and secondary dengue in Paraguay and its relation to virus serotype. *The Journal of Infection in Developing Countries, 13*(12), 1127–1134. https://doi.org/10.3855/jidc.11584

Medina, F., Medina, J. F., Colón, C., Vergne, E., Santiago, G. A., & Muñoz-Jordán, J. L. (2012). Dengue virus: Isolation, propagation, quantification, and storage. *Current Protocols in Microbiology, 27*(1). https://doi.org/10.1002/9780471729259.mc15d02s27

Peeling, R. W., Artsob, H., Pelegrino, J. L., Buchy, P., Cardosa, M. J., Devi, S., ... Yoksan, S. (2010). Evaluation of diagnostic tests: Dengue. *Nature Reviews Microbiology, 8*(S12), S30–S37. https://doi.org/10.1038/nrmicro2459

Prescott, J., Feldmann, H., & Safronetz, D. (2017). Amending Koch's postulates for viral disease: When "growth in pure culture" leads to a loss of virulence. *Antiviral Research, 137,* 1–5. https://doi.org/10.1016/j.antiviral.2016.11.002

Salles, T. S., da Encarnação Sá-Guimarães, T., de Alvarenga, E. S. L., Guimarães-Ribeiro, V., de Meneses, M. D. F., de Castro-Salles, P. F., ... Moreira, M. F. (2018). History, epidemiology and diagnostics of dengue in the American and Brazilian contexts: A review. *Parasites & Vectors, 11*(1), 264. https://doi.org/10.1186/s13071-018-2830-8

Schilling, S., Ludolfs, D., Van An, L., & Schmitz, H. (2004). Laboratory diagnosis of primary and secondary dengue infection. *Journal of Clinical Virology, 31*(3), 179–184. https://doi.org/10.1016/j.jcv.2004.03.020

Sekaran, S. D., Ew, C. L., Subramaniam, G., & Kanthesh, B. M. (2009). Sensitivity of dengue virus NS-1detection in primary and secondary infections.

African Journal of Microbiology Research. https://doi.org/10.5897/AJMR.9000647

Sekaran, S. D., Lan, E. C., & Subramaniam, G. (2018). Comparison of five serological diagnostic assays for the detection of IgM and IgG antibodies to dengue virus. *BMJ Open*, *8*(1), 1–14.

Sekaran, Shamala Devi, & Artsob, H. (2007). Molecular diagnostics for the detection of human flavivirus infections. *Expert Opinion on Medical Diagnostics*, *1*(4), 521–530. https://doi.org/10.1517/17530059.1.4.521

Sekaran, Shamala Devi, & Soe, H. J. (2017). Issues in contemporary and potential future molecular diagnostics for dengue. *Expert Review of Molecular Diagnostics*, *17*(3), 217–223. https://doi.org/10.1080/14737159.2017.1275963

Shamala Devi, S. (2008). Dengue: Breakbone fever, hemorrhagia or shock. *JUMMEC*, *11*(2), 39–52.

Skae, F. M. T. (1902). Dengue fever in penang. *BMJ*, *2*(2185), 1581–1582. https://doi.org/10.1136/bmj.2.2185.1581-a

Tricou, V., Vu, H. T., Quynh, N. V., Nguyen, C. V., Tran, H. T., Farrar, J., … Simmons, C. P. (2010). Comparison of two dengue NS1 rapid tests for sensitivity, specificity and relationship to viraemia and antibody responses. *BMC Infectious Diseases*, *10*(1), 142. https://doi.org/10.1186/1471-2334-10-142

Uno, N. & Ross, T. M. (2018). Dengue virus and the host innate immune response. *Emerging Microbes & Infections*, *7*(1), 1–11. https://doi.org/10.1038/s41426-018-0168-0

Wang, S. M. & Sekaran, S. D. (2010a). Early diagnosis of dengue infection using a commercial dengue duo rapid test kit for the detection of NS1, IgM, and IgG. *The American Journal of Tropical Medicine and Hygiene*, *83*(3), 690–695. https://doi.org/10.4269/ajtmh.2010.10-0117

Wang, S. M. & Sekaran, S. D. (2010b). Evaluation of a commercial SD dengue virus NS1 antigen capture enzyme-linked immunosorbent assay kit for early diagnosis of dengue virus infection. *Journal of Clinical Microbiology*, *48*(8), 2793–2797. https://doi.org/10.1128/JCM.02142-09

World Health Organization. (2009). Laboratory diagnosis and diagnostic tests. In *Dengue: Guidelines for Diagnosis, Treatment, Prevention and Control: New Edition*.

Chapter 2

Traditional and Current Diagnosis

Wang Seok Mui[a] and Chandramathi Samudi Raju[b]

[a]*Department of Medical Microbiology & Parasitology, Faculty of Medicine, Universiti Teknologi MARA, Selangor, Malaysia*

[b]*Department of Medical Microbiology, Faculty of Medicine, University of Malaya, Kuala Lumpur, Malaysia*

wangsm@uitm.edu.my, chandramathi@um.edu.my

Dengue virus causes a spectrum of illnesses ranging from undifferentiated mild febrile illness to severe dengue hemorrhagic fever/dengue shock syndrome. Diagnosing dengue fever can be challenging due to its signs and symptoms and can be easily confused with other diseases such as influenza, chikungunya, malaria, leptospirosis, and typhoid fever. Clinical diagnosis alone is not always reliable, as dengue has no pathognomonic clinical features from other febrile illnesses (Rathakrishnan & Sekaran, 2013). As there is no specific treatment for dengue, timely and careful clinical management of patients by experienced healthcare providers frequently saves lives. Hence, early definitive detection of infection is essential, and this can be achieved by confirmatory tests of the patient's samples. Although various laboratory methods are currently used for diagnosing dengue, none fulfill the ideal requirement for both sensitivity and specificity, while also being

Dengue Diagnostics: The Right Test at the Right Time for the Right Group
Edited by Shamala Devi Sekaran

ISBN 978-981-4968-97-3 (Hardcover), 978-1-032-66956-4 (eBook)
www.jennystanford.com

rapid, simple, and low cost. Newer detection methods for dengue that can fill this acknowledged gap are urgently needed.

Common laboratory methods for diagnosing dengue include virus isolation, detection of viral antigens, genomic detection by molecular methods, and serology. Dengue diagnosis is divided into two main phases: the early phase and the late phase. Different laboratory methods are being used to diagnose dengue at different phases of illness (Figure 2.1). In the early or febrile phase of illness, the approaches of diagnosis are via viral isolation, RT-PCR, and antigen detection, while the later phase is mainly through serology (Shamala, 2015). For a diagnosis of 'confirmed' dengue, dengue virus should be isolated or viral RNA amplified, or there should be a four-fold rise in antibody titer in paired (acute and convalescent) serum specimens, or demonstration of seroconversion from negative to positive IgM antibody. Serological tests such as IgM Capture ELISA are favorable, especially in hospitals of dengue-endemic countries due to their inexpensiveness, simplicity, and narrow time frame of viremia (Lee et al., 2011). Nevertheless, serological assays can be a challenge in hyperendemic areas where pre-existing antibodies complicate diagnosis (Sekaran et al., 2014). The choice of a test, therefore, depends not only on the availability of facilities and human resources but more importantly on the time of sampling.

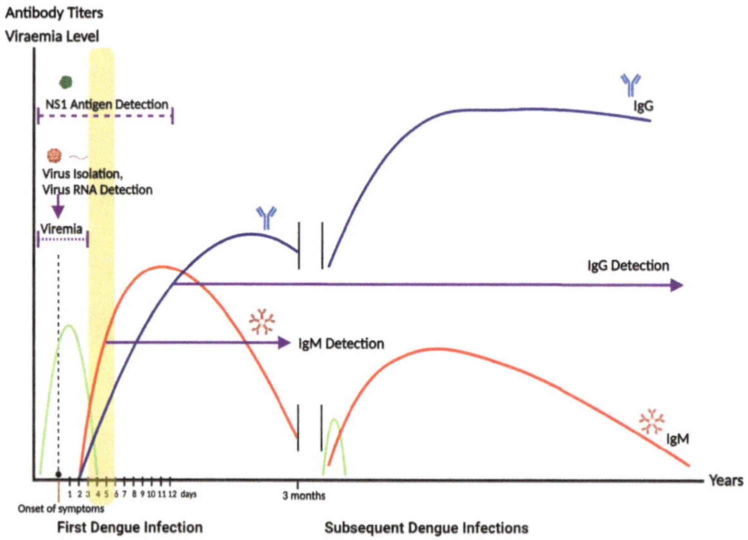

Figure 2.1 Suitability of dengue diagnostics at different phases of illness.

2.1 Virus Isolation

Virus isolation has been widely used as a gold standard in dengue diagnosis (Velez, Kuno, Gubler, Oliver, & Sather, 1984). The method involves the isolation of viruses using various approaches such as inoculation into mosquito-derived cell lines or mammalian cell lines, intrathoracic and intracerebral inoculation of adult mosquitoes, intracerebral inoculation of mosquito larvae, and suckling mice brains (De Paula & Fonseca, 2004; Guzmán & Kourí, 1996; Philip Samuel & Tyagi, 2006; Shu & Huang, 2004). Virus isolation is highly achievable when specimens are collected during the viremic phase, which is 2–3 days prior to the onset of fever, and lasts for a further 2–3 days (Vorndam & Kuno, 1997). Therefore, samples for virus isolation must be taken in the first few days of the disease. An acute phase serum sample is the specimen of choice for routine dengue virus isolation; however, dengue virus can also be detected in plasma, peripheral blood, cerebrospinal fluid, pleural fluid, and tissues of reticuloendothelial origin such as the liver, spleen, lymph nodes, lung, and thymus (Guzmán & Kourí, 2004; Vorndam & Kuno, 1997). Following harvest, confirmation tests are then performed under a UV microscope using an immunofluorescence method with dengue-specific monoclonal antibodies or detection of viral nucleic acid using a polymerase chain reaction. Thus, this method requires skill and substantial equipment to be carried out and is not the choice for many laboratories. This method depends heavily on the survival of the sample. Hence, the timing of specimen collection, proper storage, and transportation are pertinent, as these factors may affect virus viability. However, virus isolation remains very useful and relevant as a diagnostic tool, especially for monitoring dengue epidemiology and evolution as well as determining its antigenic drift (Sekaran et al., 2014).

2.1.1 Intracerebral Injection into Mice

Intracerebral inoculation of suckling mice (e.g., BALB/c mice) has been used successfully in isolating all four dengue serotypes in 1-day-old suckling using 10–20 μl of serum (Singharaj, Simasathien, Sukhavachana, Halstead, & Scanlon, 1966). These suckling mice may exhibit encephalitis symptoms (Amaral et al., 2011). It is the oldest and least sensitive method for virus isolation. This method is only used when no other methods are available. Nevertheless, it offers the advantage of evaluating the neurotropic characteristics of the dengue virus particle in vivo. Viruses are generally identified by the immunofluorescence method using dengue-specific monoclonal antibodies applied to the brain tissue of mice. Not all laboratories house animal facilities nor the ready availability of suckling. As such, storage of samples till the availability of the suckling would result in a lower sensitivity especially when viral loads are low. The advantage of using suckling mice is that other arboviruses that cause dengue-like symptoms may be isolated with this system (See Appendix 1 for protocol).

2.1.2 Inoculation of Specimen into Mosquitoes

Mosquito inoculation remains the most sensitive system for dengue virus isolation and this specific method for isolation of DENV takes between 2 and 5 days in adults and 2 and 3 days in the 4th instar larvae of *Toxorhynchites splendens* (Lam et al., 1986). In adults and larvae, intracerebral and intrathoracic inoculation have yielded comparable results with the intracerebral method being more sensitive (Amaral et al., 2011; Guzmán & Kourí, 1996; Lam et al., 1986). This method has been under-utilized because most dengue laboratories do not have insectary facilities and equipment to breed mosquitoes, and precautions need to be taken to avoid the release of infected mosquitoes. Besides, the technique requires good eye-hand coordination while doing manipulation under a stereoscopic microscope. Extensive practice is needed to become proficient at inoculating mosquitoes. These highly skilled techniques in conducting direct mosquito inoculation make in vitro cell culture a preferred option (Philip Samuel & Tyagi, 2006). Vectors for the dengue virus such as *Aedes aegypti, Aedes albopictus,* and *Aedes polynesiensis,* and non-blood feeding *Toxorhynchites* species can be used for this method. Mosquitoes of both sexes are susceptible. Generally, *Toxorhynchites* species are used because of their comparatively larger size, which facilitates the easier introduction of the inoculum (Donald, Siriyasatien, & Kohl, 2020). Besides its unique feature of not being hematophagous, it allows the work to be carried out with impunity. The drawback of this method is that between 5 and 20 mosquitoes are needed to produce results of higher sensitivity. Following inoculation, mosquitoes are held in the incubator at 32°C for 14 days to allow the dengue virus to replicate and disseminate to tissues throughout the mosquitoes (Choy, Ellis, Gubler, & Ellis, 2013). Surviving mosquitoes are then harvested and examined for the presence of viral antigens in head tissue by either immunofluorescence assay or polymerase chain reaction (See Appendix 2 for protocol).

2.1.3 Viral Culture Using Mosquito or Mammalian Cell Line

In vitro culture of cell lines is an alternative technique for isolating viruses from specimens. Because of the technical skill and special containment required for direct mosquito inoculation, cell culture is preferable for routine diagnosis, despite the greater sensitivity of using mosquitoes. The common cell lines currently utilized include the mosquito cells from *Aedes albopictus* (C6/36), *Aedes pseudoscutelaris* (AP-61, AP-64), and *Tinissa amboinensis* (TRA-284); and mammalian cells such as Vero and BS-C-1 (both derived from African green monkey kidney), rhesus monkey kidney (LLCMK2), porcine kidney (PS), and baby hamster kidney (BHK21) (Guzmán & Kourí, 1996; Igarashi, 1978). These cell lines are easily available, stable, and can be kept at room temperature for up to a week in maintenance media. Isolation of viruses typically takes 7–10 days with daily observation for cytopathic effect (CPE) before confirmation by immunofluorescence assay or polymerase chain reaction. Sometimes more than two passages may be required to isolate the virus or to increase the virus load. Not all viruses produce CPE. For these, a blind harvest may be carried out and further confirmed by confirmatory tests. Various methods have been conducted in an attempt of speeding up the isolation rate by centrifuging serum samples onto cell monolayers prior to incubation (Kao, King, Chao, Wu, & Chang, 2005; Roche, Alvarez, Guzmán, Morier, & Kourí, 2000) and locating viral antigens using flow cytometry after a few days of inoculation (Kao et al., 2001; Oliveira De Paula, Malta Lima, Clotteau, Pires Neto, & Lopes da Fonseca, 2003). Rodriguez et al. reported that 16.6% more isolates than with the conventional method were obtained when using the centrifugation technique

(Roche et al., 2000). However, these methods may not always work especially when the viral loads are low. In a comparative study, the virus isolation rate was found to be the highest in TRA-284, followed by AP-61 and C6/36 (Kuno, Gubler, Vélez, & Oliver, 1985). Nonetheless, C6/36 cells were favored for detecting infected cells by the direct fluorescent antibody test. Problems with AP-61 and TRA-284 are due to the frequent cell clumping and difficulties in forming a monolayer in a cell culture flask. For all these techniques, skilled and experienced personnel are needed to be able to recognize the CPE and the type of immunofluorescence observed upon staining with dengue-specific monoclonal antibodies (See Appendix 3 for protocol).

2.2 Detection of Dengue-Specific Antigens and Antibodies

2.2.1 Antigen Detection

Samples that could be used to detect DENV antigens during the febrile phase of dengue illness, include peripheral blood leukocytes, liver (Couvelard et al., 1999), and lung at autopsy, and less often in the thymus, lymph nodes, skin, spleen (Balsitis et al., 2009) bone marrow, and serosa (Nakao, Lai, & Young, 1989). The detection of virus antigens in frozen and paraffin-embedded tissue sections can be done using immunohistochemistry and enzyme immunoassay (EIA) techniques. These techniques include the use of labeled monoclonal antibodies that bind with detection markers such as fluorescent dyes (fluorescent antibody), enzymes (avidin-biotin enzyme and immunoperoxidase), or colloidal gold. However, the difficulty in sample collection would be the main limitation to use these detection methods as routine laboratory tests. Hence, other assay methods were developed to detect virus particles and recombinant viral proteins using NS1 and E protein (Cuzzubbo et al., 2001; Kumarasamy et al., 2007).

2.2.1.1 Non-structural protein (NS1 & NS5)-based assay

NS1 is a glycoprotein that is usually secreted during the replication of flavivirus. Infected host cells synthesize NS1, which rapidly homodimerizes in the lumen of the endoplasmic reticulum (ER) and attaches to the ER membrane. It is transported to the cell surface via the secretory pathway and released into the extracellular environment as a hexamer with lipid cargo in its middle area. This secreted NS1 (sNS1) is stable and exists at a high level in serum. It is detectable in patients' serum between day 1 and day 9 after fever onset. Hence, sNS1 can be generally used as a marker for the early diagnosis of flavivirus infections (Alcon-LePoder et al., 2008). At present, commercial assays are available only for the detection of DENV NS1 Ag. Although the assay can cross-react with NS1 antigens of other flaviviruses namely ZIKV and YFV, the assay sensitivity is high toward DENV. Also, the detection of NS1 antigen using the ELISA technique or any other rapid kit cannot differentiate between dengue serotypes.

NS5 is a potential biomarker for dengue virus infection, due to its highly conserved and immunogenic properties. In a past study, two recombinant protein products (104 kDa full-length NS5 and 70 kDa NS5-C7) were found to react with the sera of patients infected with DENV-2 suggesting its potential use as an effective antigen to detect dengue virus infection (Zhang et al., 2019). However, as most studies have focused on the detection of NS5 in only DENV-2, more data is needed to test its sensitivity toward other serotypes especially when it is used in rapid test kits.

Rapid tests are applicable in resource-limited settings where suitable healthcare facilities might not be accessible. Usually, all the rapid tests take around 30 min to perform. Nevertheless, all the rapid tests have a significant decrease in sensitivity compared to the FDA-approved Detect NS1 ELISA by InBios International. The general interpretation of the rapid test is a positive NS1 supports a DENV infection, while a negative result does not rule out infection possibility. The negative result is followed by an IgM-based test. Also, one should take note that NS1-based tests are not recommended after 7 days. There are seven commercially available dengue NS1 Ag detection assays, with four of them being rapid tests. They are Dengue NS1 Ag STRIP (BioRad, Marnes-la-Coquette, France), Platelia Dengue NS1 Ag ELISA (BioRad), Dengue NS1 Detect Rapid Test (InBios International), DENV Detect NS1 ELISA (InBios International), Panbio Dengue Early Rapid (Alere, Waltham, WA, USA), Panbio Dengue Early ELISA (2nd generation (Alere), and SD

Bioline Dengue NS1 Ag Rapid Test (Abbott, Abbott Park, IL, USA). These tests are only used for diagnostic purposes and not as point-of-care tests (Kabir, Zilouchian, Younas, & Asghar, 2021).

2.2.2 Antibody Detection

DENV-specific IgM antibodies are present in patient serum from 4 to 5 days of primary infection and are detectable for up to 3 months. IgG antibodies in serum can be seen about a week after the onset of fever, which remains at high levels for several weeks before declining. Usually, IgG can be detected for decades. Memory B cells are stimulated in response to secondary DENV infection for secreting DENV-specific IgG that is detectable even on the first day of symptoms. The key point is, the IgG titer remains much higher (more than two folds) compared to its level during primary infection. In secondary dengue, the IgM response is variable and sometimes may be undetectable. The most commonly used serology detection methods for dengue diagnosis are hemagglutination inhibition (HI) and ELISA for IgM or IgG (Kao et al., 2005).

2.2.2.1 Hemagglutination inhibition test

The HI assay is generally used to titrate the antibody response to a specific viral infection in vitro. Some viruses have the ability to hemagglutinate (bind) red blood cells and avoiding the red blood cells from clumping. The basic principle of HI assay is, if antibodies that cross-react with the virus are present in the infected person's sera sample, the antibodies will bind to the virus thus preventing the virus from hemagglutinating the red blood cells. Using HI, the exact titers of the antibodies in the sera can be determined. Primary and secondary DENV infections can be differentiated using HI but not an early infection. However, for secondary infections, the IgG-based test is superior to the HI. The disadvantage of this assay is that it may not be able to detect antibodies of other dengue virus subtypes. Also, it will not be able to detect antibodies against other structures of the virus but only the ones against the hemagglutinating protein. For this, plaque reduction neutralization tests (PRNT) will be the option (Bourgeois & Oaks, 2014; Fry et al., 2011).

The HI test (Figure 2.2) is based on the ability of dengue antigens to agglutinate red blood cells (RBC) of ganders or trypsinized human O RBC.

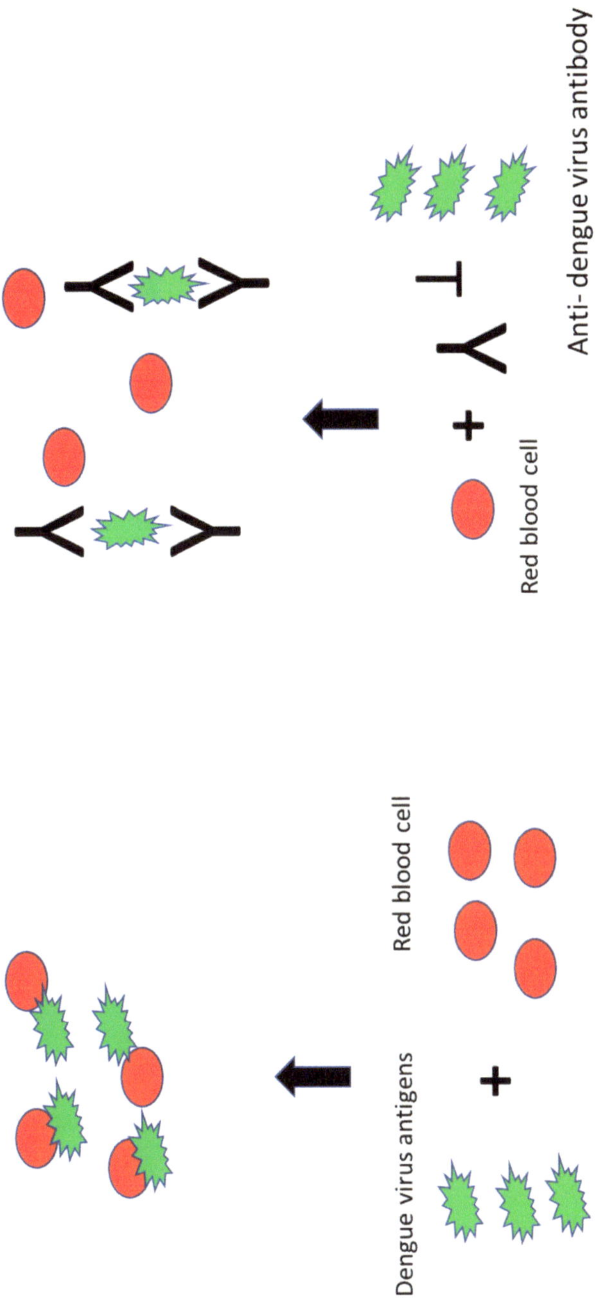

Figure 2.2 Principle of HI test.

2.2.2.2 Envelope/membrane (E/M)-antigen ELISA (both IgM and IgG)

The immune system produces antibodies to fight against an antigen. As part of the adaptive immune response towards the antigen, T cell will trigger B cells to produce immunoglobulin (Ig) proteins or antibodies (IgD, IgM, IgG, IgA, Ig E). IgM antibodies are commonly produced 5 days post onset of symptoms, and it can last for about 2–3 months and at times even longer. During dengue primary infection, IgM level is raised while raised level of IgG is seen during secondary infection (Hunsperger et al., 2009).

Various enzyme-linked immunosorbent assay capturing the IgM antibody (MAC-ELISA) testing kits that are developed and marketed for testing dengue infection globally. The basic principle of the test is that the anti-human IgM antibodies on a solid phase will capture the dengue IgM antibody in patient's serum. Apart from MAC-ELISA, a sandwich-type immunoassay has also been developed. This technique uses the recombinant dengue-derived antigen and DENV-specific monoclonal antibody. One of FDA approved kit would be the DENV Detecta IgM Capture ELISA by InBios International, Inc. which uses a 96 well plate to detect the human Dengue-IgM antibody (Namekar et al., 2013). After IgM, IgG will then be produced and this antibody can last for many years or even lifelong.

In recent secondary dengue infection, a four-fold spike in seen in the IgG level. Dengue IgG ELISA kits are developed and commercialised by Standard Diagnostics (Seoul, South Korea) and Panbio Inc (Brisbane, Australia now Alere Inc., Waltham, WA, USA) (Blacksell et al., 2012). It is known that dengue IgG antibody cross-reacts with IgG antibodies of other flaviviruses. This leads to difficulty in determining primary or secondary infection by depending on IgG level. To overcome this, IgM/IgG ratio is suggested to be used for differentiating the primary from secondary infection. A ratio between the optical density (OD) measured from the respective ELISA tests of IgM and IgG is measured whereby a ratio more than 1.32 implicates primary infection and below 1.32 implicates secondary infection (Prince et al., 2011).

2.2.2.3 Rapid diagnostic tests

Immunochromatographic tests are used for the rapid diagnosis of dengue. Such test is used to detect dengue NS1 antigen within 5 to 7 days of fever/symptom onset. A rapid test approximately takes about 30 minutes and very suitable for resource-limited settings. There is also antibody-capture immunochromatographic test cassette available for the detection of IgM and IgG which are basically to view the bands that are formed after adding a patient's whole blood, serum or plasma. Commonly used RDTs are Duo IgM and IgG Rapid Test Strip (Panbio); Bioline Dengue IgG/IgM (Standard Diagnostics); VScan (Minerava); Smartcheck (GlobaleMed); Denguecheck-WB (Tulip); and Dengue IgG/IgM (Core). Note that the Bioline™ Dengue Duo kit is a rapid, immunochromatographic assay which detects both dengue virus NS1 antigen and IgG/IgM antibodies. As the NS1 rapid test kits' sensitivity and specificity ranges (due to cross-reactivity with other flaviviruses) among various manufacturers, the dengue NS1 antigen is detected in combination with anti-glycoprotein E IgM and IgG antibodies (Fry et al., 2011).

2.3 Viral Nucleic Acid Detection

Nucleic acid tests (NATs) are molecular tests that require costly equipment (PCR machine) and skilled personnel. Dengue NATs require viral RNA extraction from patient serum/plasma, which in most cases, difficult to be performed in POC settings. NATs quantify viral RNA from patient samples and this allows detection of dengue during acute phase with higher accuracy and sensitivity. This test is applicable for five days of symptom onset. Commonly used methods are reverse transcription polymerase chain reaction (RT-PCR), nucleic acid sequence-based amplification (NASBA), and transcription-mediated amplification (TMA).

2.3.1 Polymerase Chain Reaction (PCR)

The most common DENV NAT test is PCR. In the one step reverse transcriptase-polymerase chain reaction PCR (RT-PCR), the viral RNA extracted from different patient samples (e.g., plasma, blood, urine, and serum) will be added with primer pairs and probes which are specific to each dengue serotypes (DENV1- DENV4). The use of fluorescent probe (e.g., SYBR green, Taqman®) enables the detection and quantification (qRT-PCR) of the amplified product in real time using PCR machine. Generally, compared to the SYBR green, the TaqMan real-time PCR is highly specific as it uses the sequence-specific hybridization of the probe. The available qRT-PCR include the "singleplex" (detecting only one serotype) or "multiplex" (able to identify all four serotypes in each single sample). A multiplex RT-PCR assay using SYBR green probe was developed to determine all four serotypes from the blood sample (Yong et al., 2007). The developed assay had a sensitivity and specificity of 98.18% and 100%, respectively. Another multiplex one-step RT-PCR assay has been developed using TaqMan probe (see Appendix 4 for protocol) (Kong et al., 2006). The assay uses a set of forward and reverse primers designed to target serotype conserved region at the NS5 gene, and simultaneously target a variable region for all four serotypes (sets of primers that were previously used to design the serotype-specific TaqMan probes). This assay had sensitivity, specificity, and real-time PCR efficiency of 89.54%, 100%, and 91.5%, respectively. Currently, there are two NAT assays approved by the FDA: CDC DENV-1-4 Real-Time RT-PCR Multiplex assay and Trioplex rRT-PCR Assay. The former (differentiates all four serotypes) is used in cases where the primary infection cause would be the DENV. The Trioplex rRT-PCR Assays are used when Dengue, Chikungunya, and Zika are suspected, allowing one to screen for all three viruses at the same time (Kabir et al., 2021).

2.3.2 Nucleic Acid Sequence–Based Amplification (NASBA)

NASBA is a one-step isothermal process for amplifying RNA that does not require the use of a thermal cycler. This technique has been proven to be successful in the detection of both viral and bacterial RNA in clinical samples. The NASBA assay involves the use of extracted nucleic acid and subsequently amplified at a constant temperature of 41°C. The reaction mixture consists of avian myeloblastosis reverse transcriptase (AMV-RT), T7 RNA polymerase, and RNase H with two short single-stranded DNA primers. The amplified product of RNA pools is then detected with agarose gel electrophoresis and electrochemiluminescence (ECL) signal count. NASBA has significant specificity and sensitivity, comparable with that of virus isolation, and may be particularly useful in field studies of dengue infection. Wu et al. reported that the NASBA assay had a sensitivity of 98.5% (66/67 clinical samples), and a specificity of 100% (21 normal human serum) when compared with the cell culture method using C6/36 cells (Wu et al., 2001). The effectiveness of this method has also been described for detecting the dengue virus within mosquitoes (Jittmittraphap et al., 2006). It is considered more efficient than PCR in that it bypasses the use of a thermal cycler, which makes it both rapid (<1 day) and cost-effective (Usawattanakul & P, 2002) (See Appendix 5 for protocol).

2.3.3 Loop-Mediated Isothermal Amplification (LAMP)

The reverse transcriptase (RT-LAMP) assay was first developed by Notomi et al. (2000). The assay consists of four to six primers (2 outer primes, 2 inner primers, and 2 loop primers) which are able to recognize eight distinct regions on the target. This method needs only water bath or a simple heating block to amplify the viral target at a persistent temperature (range of 60–65°C). The assay starts with transcription of the target to DNA, followed by creating a loop structure which generates target copies. The amplified target is detected by the naked eye or sometimes under UV light.

The RT-LAMP assay is previously known to have 100% and 98.9% success rate in detecting the virus in the clinical DENV strains and infected patients, respectively compared to the RT-PCR that shows a sensitivity of 93% for clinical strains and 84.2% for infected patients. It has no false positives results (Hu et al., 2015). Therefore, RT-LAMP has many advantages as it is rapid, cost-effective, isothermal, highly sensitive, and specific. Nevertheless, PCR is superior to LAMP-based assays as it is suitable for multiplexing and viral quantification purposes. Usually, LAMP-based assay provides qualitative (detected/not detected) results.

2.4 Diagnostic Limitation

Ideally, a laboratory test for the detection of dengue should be rapid, simple, affordable, with high sensitivity and specificity, and able to detect dengue at any phase of illness, preferably able to distinguish primary and secondary dengue infections, as well as the different serotypes. Therefore, the ideal diagnostic test should be able to detect genuine dengue cases at any stage of the illness. The challenges in developing ideal diagnostic tools have been hindered by the (i) complexity of dengue pathogenesis; (ii) hyperendemicity and multiple sequential infections; and (iii) clinical conditions of patients including viremia and antibody response (Sekaran et al., 2014). Although virus isolation is more specific, it is restricted by being time-consuming, is expensive due to setting up and maintenance of special infrastructures such as animal care facilities, insectary, and cell culture, requiring expertise and skilled personnel, needing good quality acute specimens, and the inability to differentiate between primary and secondary infections (Guzmán & Kourí, 2004). Molecular detection of viral nucleic acid allows the detection of concurrent infections by multiple serotypes and determining strain variability with high sensitivity and specificity within a short time. Despite many genomic detection methods for dengue detection, not all have been evaluated and in one such external assurance study, only 10.9% of participating laboratories with different molecular-based methods met all criteria for optimal performance (Guzman et al., 2010). Likewise, the technique is also limited by the need for acute samples, specialized equipment, and skilled personnel, and by its incapacity to distinguish between primary from secondary infections (Clarke & Casals, 1958). While these limitations are being overcome by the detection of NS1 antigen, its sensitivity compared to genome detection methods is poor and cannot be guaranteed (Kassim, Izati, TgRogayah, Apandi, & Saat, 2011). Nevertheless, this antigen detection method is useful for low-resource settings, as the kits are affordable and simple to use (Tricou et al., 2010). Serological diagnosis has been the mainstay for dengue diagnosis; however, IgM detection is only detectable by 4–9 days in primary infections and by 3–5 days in secondary infections. IgG tends to cross-react with other flaviviruses (Peeling et al., 2010). Combinations of different methods such as the detection of both NS1 and IgM/IgG have proved to increase the overall performance (Blacksell et al., 2011). Hence, it seems that the best way currently to diagnose and confirm dengue is to run multiple assays or to obtain paired sera.

2.5 Future Directions

Overall, the development of diagnostic tool which is highly sensitive and specific involves multiple challenges that need be addressed. Combining the existing techniques could be one way of improving both sensitivity and specificity of diagnosing dengue. Existing SD BIOLINE Dengue Duo RDT (Abbott, Santa Clara, USA; former Alere Inc, Waltham, USA) that could detect all three NS1, dengue IgM and IgG is known to have good specificities, but low sensitivity, indicating that it could be useful detecting rather than ruling out a dengue diagnosis especially in endemic areas (Kikuti et al., 2019).

Therefore, numerous researchers have begun to focus on the use of biosensors as effective tool for diagnosing infectious disease(s) (Bhalla et al., 2016). A biosensor developed based on silicon nanowire could detect the 'reverse transcription polymerase chain reaction' product of a sample within 30 min (Zhang et al., 2010). Sensitive and specific results obtained from this system could counterpoise the other labor-intensive and time-consuming laboratory techniques namely ELISA, that usually requires well-trained laboratory staffs to carry out the test. In dengue diagnosis, a biosensors-based semi-quantitative anti-dengue IgG (immunoglobulin G) immuno-magnetic agglutination assay for the diagnosis of both previous and recent dengue infection in a single test has been developed. The test's good point-of-care diagnostic accuracy is suggested to ease the conduct of dengue serosurveys for vaccination strategy and facilitate pre-vaccination screening to ensure safety (Chong et al., 2021). The biosensor technology can also be applied as point-of-care test (POCT).

POCT technologies should be generally cheaper and reliable in health care settings. Examples of such tests are electrochemical impedance spectroscopy (EIS) based sensing, microarray-based sensing, and surface plasmon resonance (SPR). Most recently, nanoparticles are often combined with these technologies to enhance their performance in diagnosing dengue in timely manner which is essential for patient management. In addition, development of all assays should utilise respective controls and proficiency testing in order to warrant a high degree of assurance.

References

Alcon-LePoder, S., Sivard, P., Drouet, M.-T., Talarmin, A., Rice, C., & Flamand, M. (2008). *Secretion of Flaviviral Non-Structural Protein NS1: From Diagnosis to Pathogenesis.* pp. 233–250. https://doi.org/10.1002/0470058005.ch17

Amaral, D. C., Rachid, M. A., Vilela, M. C., Campos, R. D., Ferreira, G. P., Rodrigues, D. H., ..., & Teixeira, A. L. (2011). Intracerebral infection with dengue-3 virus induces meningoencephalitis and behavioral changes that precede lethality in mice. *Journal of Neuroinflammation, 8*(1), 23. https://doi.org/10.1186/1742-2094-8-23

Balsitis, S. J., Coloma, J., Castro, G., Alava, A., Flores, D., McKerrow, J. H., ..., & Harris, E. (2009). Tropism of dengue virus in mice and humans defined by viral nonstructural protein 3-specific immunostaining. *The American Journal of Tropical Medicine and Hygiene, 80*(3), 416–424

Bhalla, N., Jolly, P., Formisano, N., & Estrela, P. (2016). Introduction to biosensors. *Essays in Biochemistry,* 60(1), 1–8. https://doi.org/10.1042/EBC2015000

Blacksell, S. D., Jarman, R. G., Gibbons, R. V., Tanganuchitcharnchai, A., Mammen, M. P., Jr, Nisalak, A., Kalayanarooj, S., Bailey, M. S., Premaratna, R., de Silva, H. J., Day, N. P., & Lalloo, D. G. (2012). Comparison of seven commercial antigen and antibody enzyme-linked immunosorbent assays for detection of acute dengue infection. *Clinical and Vaccine Immunology: CVI, 19*(5), 804–810. https://doi.org/10.1128/CVI.05717-11

Bourgeois, M. A., & Oaks, J. L. (2014). Laboratory diagnosis of viral infections. In *Equine Infectious Diseases.* pp. 132–140.e2. Elsevier. https://doi.org/10.1016/B978-1-4557-0891-8.00012-9

Chong, Z. L., Soe, H. J., Ismail, A. A., Mahboob, T., Chandramathi, S., & Sekaran, S. D. (2021). Evaluation of the diagnostic accuracy of a new biosensors-based rapid diagnostic test for the point-of-care diagnosis of previous and recent dengue infections in Malaysia. *Biosensors, 11,* 129. https://doi.org/10.3390/bios11050129

Choy, M. M., Ellis, B. R., Gubler, D. J., & Ellis, E. M. (2013). Comparison of the mosquito inoculation technique and quantitative real time polymerase chain reaction to measure dengue virus concentration. *The American*

Journal of Tropical Medicine and Hygiene, 89(5), 1001–1005. https://doi.org/10.4269/ajtmh.13-0100

Clarke, D. H., & Casals, J. (1958). Techniques for hemagglutination and hemagglutination-inhibition with arthropod-borne viruses. *The American Journal of Tropical Medicine and Hygiene, 7*(5), 561–573. https://doi.org/10.4269/ajtmh.1958.7.561

Couvelard, A., Marianneau, P., Bedel, C., Drouet, M.-T., Vachon, F., Hénin, D., & Deubel, V. (1999). Report of a fatal case of dengue infection with hepatitis: Demonstration of dengue antigens in hepatocytes and liver apoptosis. *Human Pathology, 30*(9), 1106–1110. https://doi.org/10.1016/S0046-8177(99)90230-7

Cuzzubbo, A. J., Endy, T. P., Nisalak, A., Kalayanarooj, S., Vaughn, D. W., Ogata, S. A., ..., & Devine, P. L. (2001). Use of recombinant envelope proteins for serological diagnosis of dengue virus infection in an immunochromatographic assay. *Clinical Diagnostic Laboratory Immunology, 8*(6), 1150–1155. https://doi.org/10.1128/CDLI.8.6.1150-1155.2001

De Paula, S. O., & Fonseca, B. A. L. da. (2004). Dengue: A review of the laboratory tests a clinician must know to achieve a correct diagnosis. *Brazilian Journal of Infectious Diseases, 8*(6). https://doi.org/10.1590/S1413-86702004000600002

Donald, C. L., Siriyasatien, P., & Kohl, A. (2020). Toxorhynchites species: A review of current knowledge. *Insects, 11*(11), 747. https://doi.org/10.3390/insects11110747

Fry, S. R., Meyer, M., Semple, M. G., Simmons, C. P., Sekaran, S. D., Huang, J. X., ..., & Cooper, M. A. (2011). The diagnostic sensitivity of dengue rapid test assays is significantly enhanced by using a combined antigen and antibody testing approach. *PLoS Neglected Tropical Diseases, 5*(6), e1199. https://doi.org/10.1371/journal.pntd.0001199

Guzman, M. G., Halstead, S. B., Artsob, H., Buchy, P., Farrar, J., Gubler, D. J., ..., & Peeling, R. W. (2010). Dengue: A continuing global threat. *Nature Reviews Microbiology, 8*(S12), S7–S16. https://doi.org/10.1038/nrmicro2460

Guzmán, M. G., & Kourí, G. (1996). Advances in dengue diagnosis. *Clinical Diagnostic Laboratory Immunology, 3*(6), 621–627. https://doi.org/10.1128/cdli.3.6.621-627.1996

Guzmán, M. G., & Kourí, G. (2004). Dengue diagnosis, advances and challenges. *International Journal of Infectious Diseases, 8*(2), 69–80. https://doi.org/10.1016/j.ijid.2003.03.003

Henchal, E. A., & Putnak, J. R. (1990). The dengue viruses. *Clinical Microbiology Reviews, 3*(4), 376–396. https://doi.org/10.1128/CMR.3.4.376

Hunsperger, E. A., Yoksan, S., Buchy, P., Nguyen, V. C., Sekaran, S. D., Enria, D. A., Pelegrino, J. L., Vázquez, S., Artsob, H., Drebot, M., Gubler, D. J., Halstead, S. B., Guzmán, M. G., Margolis, H. S., Nathanson, C. M., Rizzo Lic, N. R., Bessoff, K. E., Kliks, S., & Peeling, R. W. (2009). Evaluation of commercially available anti-dengue virus immunoglobulin M tests. *Emerging infectious diseases, 15*(3), 436–440. https://doi.org/10.3201/eid1503.080923

Igarashi, A. (1978). Isolation of a Singh's Aedes albopictus cell clone sensitive to dengue and chikungunya viruses. *Journal of General Virology, 40*(3), 531–544. https://doi.org/10.1099/0022-1317-40-3-531

Jittmittraphap, A., Thammapalo, S., Ratanasetyuth, N., Wongba, N., Mammen, M. P., & Jampangern, W. (2006). Rapid detection of dengue viral RNA in mosquitoes by nucleic acid-sequence based amplification (NASBA). *The Southeast Asian Journal of Tropical Medicine and Public Health, 37*(6), 1117–1124.

Kabir, M. A., Zilouchian, H., Younas, M. A., & Asghar, W. (2021). Dengue detection: Advances in diagnostic tools from conventional technology to point of care. *Biosensors, 11*(7), 206. https://doi.org/10.3390/bios11070206

Kao, C.-L., King, C.-C., Chao, D.-Y., Wu, H.-L., & Chang, G.-J. J. (2005). Laboratory diagnosis of dengue virus infection: current and future perspectives in clinical diagnosis and public health. *Journal of Microbiology, Immunology, and Infection, 38*(1), 5–16.

Kao, C.-L., Wu, M.-C., Chiu, Y.-H., Lin, J.-L., Wu, Y.-C., Yueh, Y.-Y., ..., & King, C.-C. (2001). Flow cytometry compared with indirect immunofluorescence for rapid detection of dengue virus type 1 after amplification in tissue culture. *Journal of Clinical Microbiology, 39*(10), 3672–3677. https://doi.org/10.1128/JCM.39.10.3672-3677.2001

Kassim, F. M., Izati, M. N., TgRogayah, T. A. R., Apandi, Y. M., & Saat, Z. (2011). Use of dengue NS1 antigen for early diagnosis of dengue virus infection. *The Southeast Asian Journal of Tropical Medicine and Public Health, 42*(3), 562–569.

Kikuti, M., Cruz, J. S., Rodrigues, M. S., Tavares, A. S., Paploski, I. A. D., Silva, M. M. O., Santana, P. M., Tauro, L. B., Silva, G. A. O. F., Campos, G. S., Araújo, J. M. G., Kitron, U., Reis, M. G., & Ribeiro, G. S. (2019). Accuracy of the SD BIOLINE Dengue Duo for rapid point-of-care diagnosis of dengue. *PloS One, 14*(3), e0213301. https://doi.org/10.1371/journal.pone.0213301

Kong, Y. Y., Thay, C. H., Tin, T. C., & Devi, S. (2006). Rapid detection, serotyping and quantitation of dengue viruses by TaqMan real-time one-step RT-PCR. *Journal of Virological Methods, 138*(1–2), 123–130. https://doi.org/10.1016/j.jviromet.2006.08.003

Kumarasamy, V., Wahab, A. H. A., Chua, S. K., Hassan, Z., Chem, Y. K., Mohamad, M., & Chua, K. B. (2007). Evaluation of a commercial dengue NS1 antigen-capture ELISA for laboratory diagnosis of acute dengue virus infection. *Journal of Virological Methods, 140*(1–2), 75–79. https://doi.org/10.1016/j.jviromet.2006.11.001

Kuno, G., Gubler, D. J., Vélez, M., & Oliver, A. (1985). Comparative sensitivity of three mosquito cell lines for isolation of dengue viruses. *Bulletin of the World Health Organization, 63*(2), 279–286.

Lam, S. K., Chew, C. B., Poon, G. K., Ramalingam, S., Seow, S. C., & Pang, T. (1986). Isolation of dengue viruses by intracerebral inoculation of mosquito larvae. *Journal of Virological Methods, 14*(2), 133–140. https://doi.org/10.1016/0166-0934(86)90044-3

Lee, L. K., Thein, T. L., Kurukularatne, C., Gan, V. C., Lye, D. C., & Leo, Y. S. (2011). Dengue knowledge, attitudes, and practices among primary care physicians in Singapore. *Annals of the Academy of Medicine, Singapore, 40*(12), 533–538.

Namekar, M., Ellis, E. M., O'Connell, M., Elm, J., Gurary, A., Park, S. Y., Imrie, A., & Nerurkar, V. R. (2013). Evaluation of a new commercially available immunoglobulin M capture enzyme-linked immunosorbent assay for diagnosis of dengue virus infection. *Journal of Clinical Microbiology, 51*(9), 3102–3106. https://doi.org/10.1128/JCM.00351-13

Nakao, S., Lai, C. J., & Young, N. S. (1989). Dengue virus, a flavivirus, propagates in human bone marrow progenitors and hematopoietic cell lines. *Blood, 74*(4), 1235–1240.

Notomi, T. (2000). Loop-mediated isothermal amplification of DNA. *Nucleic Acids Research, 28*(12), 63e–63. https://doi.org/10.1093/nar/28.12.e63

Oliveira De Paula, S., Malta Lima, D., Clotteau, M., Pires Neto, R. da J., & Lopes da Fonseca, B. A. (2003). Improved detection of dengue-1 virus from IgM-positive serum samples using C6/36 cell cultures in association with RT-PCR. *Intervirology, 46*(4), 227–231. https://doi.org/10.1159/000072432

Peeling, R. W., Artsob, H., Pelegrino, J. L., Buchy, P., Cardosa, M. J., Devi, S., ..., & Yoksan, S. (2010). Evaluation of diagnostic tests: Dengue. *Nature Reviews Microbiology, 8*(S12), S30–S37. https://doi.org/10.1038/nrmicro2459

Philip Samuel, P., & Tyagi, B. K. (2006). Diagnostic methods for detection & isolation of dengue viruses from vector mosquitoes. *The Indian Journal of Medical Research, 123*(5), 615–628.

Prince, H. E., Yeh, C., & Lapé-Nixon, M. (2011). Utility of IgM/IgG ratio and IgG avidity for distinguishing primary and secondary dengue virus infections using sera collected more than 30 days after disease onset. *Clinical and Vaccine Immunology: CVI, 18*(11), 1951–1956. https://doi.org/10.1128/CVI.05278-11

Rathakrishnan, A., & Sekaran, S. D. (2013). New development in the diagnosis of dengue infections. *Expert Opinion on Medical Diagnostics, 7*(1), 99–112. https://doi.org/10.1517/17530059.2012.718759

Roche, R. R., Alvarez, M., Guzmán, M. G., Morier, L., & Kourí, G. (2000). Comparison of rapid centrifugation assay with conventional tissue culture method for isolation of dengue 2 virus in C6/36-HT cells. *Journal of Clinical Microbiology, 38*(9), 3508–3510. https://doi.org/10.1128/JCM.38.9.3508-3510.2000

Sekaran, S. D., Rathakrishnan, A., & Yeo, A. S. L. (2014). Dengue: An overview. *Journal of Health and Translational Medicine, 17*(2), 1–10. https://doi.org/10.22452/jummec.vol17no2.1

Shamala, D. S. (2015). Laboratory diagnosis of dengue: A review. *International Medical Journal Malaysia, 14*(1), 17–28. https://doi.org/10.31436/imjm.v14i1.452

Shu, P.-Y., & Huang, J.-H. (2004). Current advances in dengue diagnosis. *Clinical and Vaccine Immunology, 11*(4), 642–650. https://doi.org/10.1128/ CDLI.11.4.642-650.2004

Singharaj, P., Simasathien, P., Sukhavachana, P., Halstead, S. B., & Scanlon, J. E. (1966). Recovery of dengue and other viruses in mice and tissue culture from Thai mosquitos. *Bulletin of the World Health Organization, 35*(1), 67–68.

Tricou, V., Vu, H. T., Quynh, N. V., Nguyen, C. V., Tran, H. T., Farrar, J., ..., & Simmons, C. P. (2010). Comparison of two dengue NS1 rapid tests for sensitivity, specificity and relationship to viraemia and antibody responses. *BMC Infectious Diseases, 10*(1), 142. https://doi.org/10.1186/1471-2334-10-142

Usawattanakul, W., & P, A. J. T. (2002). Rapid detection of dengue viral RNA by nucleic acid sequence-based amplification (NASBA), *26*(2), 125–130.

Velez, M., Kuno, G., Gubler, D. J., Oliver, A., & Sather, G. E. (1984). Mosquito cell cultures and specific monoclonal antibodies in surveillance for dengue viruses. *The American Journal of Tropical Medicine and Hygiene, 33*(1), 158–165. https://doi.org/10.4269/ajtmh.1984.33.158

Vorndam, V., & Kuno, G. (1997). *Laboratory Diagnosis of Dengue Virus Infections Dengue and Dengue Hemorrhagic Fever*. (Gubler, D. J., & Kuno, G., Eds.). London, United Kingdom: Dengue and Dengue Hemorrhagic Fever.

Wu, S.-J. L., Lee, E. M., Putvatana, R., Shurtliff, R. N., Porter, K. R., Suharyono, W., ..., & Romano, J. W. (2001). Detection of dengue viral RNA using a nucleic acid sequence-based amplification assay. *Journal of Clinical Microbiology, 39*(8), 2794–2798. https://doi.org/10.1128/JCM.39.8.2794-2798.2001

Yong, Y. K., Thayan, R., Chong, H. T., Tan, C. T., & Sekaran, S. D. (2007). Rapid detection and serotyping of dengue virus by multiplex RT-PCR and real-time SYBR green RT-PCR. *Singapore Medical Journal, 48*(7), 662–668.

Zhang, G. J., Luo, Z. H., Huang, M. J., Tay, G. K., & Lim, E. J. (2010). Morpholino-functionalized silicon nanowire biosensor for sequence-specific label-free detection of DNA. *Biosensors & Bioelectronics, 25*(11), 2447–2453. https://doi.org/10.1016/j.bios.2010.04.001

Zhang, T., Wang, M.-L., Zhang, G.-R., Liu, W., Xiao, X.-Q., Yang, Y.-S., ..., & Chan, P. K. S. (2019). Recombinant DENV 2 NS5: An effective antigen for diagnosis of DENV infection. *Journal of Virological Methods, 265*, 35–41. https:// doi.org/10.1016/j.jviromet.2018.12.005

Chapter 3

Newer Diagnostics for Dengue Disease

Crystale Lim Siew Ying, Nur Haliza Hassan, and Shamala Devi Sekaran
Faculty of Applied Sciences, UCSI University Kuala Lumpur Campus, Kuala Lumpur, Malaysia
shamala@ucsiuniversity.edu.my

3.1 Biosensors in Dengue Diagnosis

With all vector-borne diseases, the success of dengue surveillance programs is determined by the speed and accuracy of diagnosis, more so due to the lack of specific antiviral treatments for dengue. Diagnosis in the laboratory (confirmatory testing) typically depends on direct detection methods of virus isolation, genome detection, and antigen detection, or indirect detection methods of serology IgG or IgM. Other than virus isolation, the other methods usually utilize ELISAs and reverse transcription real-time PCR (RT-qPCR). These tests are generally not able to be carried out by the general public and require specialized equipment, facilities, and expertise, both for the technical and analytical aspects.

Dengue Diagnostics: The Right Test at the Right Time for the Right Group
Edited by Shamala Devi Sekaran

ISBN 978-981-4968-97-3 (Hardcover), 978-1-032-66956-4 (eBook)
www.jennystanford.com

Biosensor technology can circumvent the limitations above while retaining almost all the advantages of laboratory-based diagnostics. Biosensors are analytical devices that involve integration between a molecular recognition platform with a physicochemical transducer to produce a single detection processing unit (Miserere & Merkoçi, 2015). In simpler terms, a biosensor is "a device that measures biological or chemical reactions by generating signals proportional to the concentration of an analyte in the reaction" (Bhalla et al., 2016). In healthcare applications, lateral flow pregnancy tests and glucose monitoring sensors are two of the most well-known and well-used biosensor devices (Bhalla et al., 2016), with the most recent success story being Covid-19 antigen rapid test kit (RTK-Ag).

The typical biosensor flow begins with the recognition or binding of the analyte (the target of detection) to the bioreceptor (a molecule specific to the analyte). This interaction between the analyte and bioreceptor is termed "biorecognition", which generates a signal in the form of light, heat, pH, charge, or mass change (Bhalla et al., 2016). The transducer then quantifies this biorecognition signal and converts it into a measurable signal in a concentration-dependent manner in a process called "signalization" via an electrical interface. Electronics recognize, convert, and amplify the analog signal to a digital one (signal conditioning) and process the signal for display as an output signal, which can be in the form of numerical, graphical, imagery, and other types of data. The biosensor workflow and options are summarized in Table 3.1.

The suitability of biosensors is evaluated by a set of parameters based on the International Union of Pure and Applied Chemistry (IUPAC) guidelines, which include specificity, selectivity, limit of detection (LOD), and sensitivity (den Boef & Hulanicki, 1983; Persson, 2001). LOD, a quantifiably accurate amount of element derived from the smallest signal, is usually calculated as $3s/b$, with s as the standard deviation of the signal of the blank sample and b as the slope of the straight section of the calibration curve (Long & Winefordner, 1983). The signal-to-noise ratio ($S/N = 3$) is an alternative LOD calculation (Shrivastava & Gupta, 2011).

Table 3.1 Biosensor workflow and options

Workflow	**Option**					
Analyte	Virus antigen protein	Whole virus	Complementary virus DNA	Virus-specific antibody		
Bioreceptor	Enzymes/ peptides	Cells	Nucleic acids/ peptide nucleic acids	Nanoparticles/ aptamers	Antibodies	Whole virus
Biorecognition method	Light	Heat	Change in pH	Change in mass		
			Transducer signal			
Transducer	Photodiode	Thermistor	pH electrode	Quartz electrode		
			Electronics Conversion from analog to digital signals Signal conditioning			
			Quantification of processed signal			
			Information display			

Biosensors may be classified according to the bioreceptor or the transducer in use (Table 3.2). Biosensors are also differentiated further based on their bioreceptors, with the options of enzymes, antibodies, nucleic acids, cells, or biomimetics. Alternatively, biosensors can also be categorized by their transducers as electrochemical, physicochemical, optical, piezoelectric, or a combination of these. Electrochemical transducers produce amperometric/voltammetric (change of current), conductometric (change of conductance), impedance (change in impedance), or potentiometric (change of membrane potential) signals, which form the mechanism of label-free electrochemical biosensors, while optical transducers produce chemiluminescence, fluorescence, or surface plasmon resonance signals.

Table 3.2 Categorization of biosensors

Bioreceptor	Transducer		
	Electrochemical	**Optical**	**Piezoelectric**
Enzyme	Amperometric	Surface plasmon resonance	
Antibody	Conductometric	Chemiluminescence	
DNA/RNA	Impedance	Fluorescence	
Cell	Potentiometric		
Biomimetics			

Electrochemical biosensors for dengue detection have become the overwhelming method of choice among researchers due to their ability to be highly quantitative and qualitative. Much attention is placed on the transducer surface selection, which immobilizes the biorecognition element. The biorecognition element is usually immobilized on a self-assembling monolayer or a thin tethering layer of polymers, where this layer provides the functional groups for the chemical reaction to occur (Goode, Rushworth, & Millner, 2015). The biorecognition elements are mainly immobilized into transducers by methods of attachment such as adsorption, crosslinking, entrapment, and microencapsulation, where covalent binding is desirable due to the bond stability (Bazin, Tria, Hayat, & Marty, 2017).

Among the electrochemical methods, the electrochemical impedance spectroscopy (EIS) approach has been increasing in popularity due to the reusability of the electrode material and the low requirement of chemical reagents (Lisdat & Schäfer, 2008). The EIS principle of analysis relies on the detection of the difference in electrode resistance before and after the interaction of the bioreceptor with its analyte (Faria & Zucolotto, 2019).

The detection of viral load in patients is still considered the gold standard for early-stage active dengue infection diagnosis as well as for patient monitoring although various tools have been developed to detect dengue from the onset of dengue symptoms to 10 days post-infection. Thus, an ideal dengue diagnostic tool should be economical, rapid, and easy to use, coupled with high sensitivity and specificity (Shamala, 2015). The hunt goes on for such a tool that is also able to differentiate between primary and secondary infections as well as serotypes (Shamala, 2015). Many efforts have not culminated in the goal of creating a single ideal assay for the diagnosis of dengue.

The typical DNA/RNA biosensor system to detect DENV includes reverse transcription real-time PCR that is based on the amplification of viral nucleic acid. Among the molecular biosensor tools, a promising point-of-care (POC) test for the rapid detection of the four DENV serotypes is found in the form of a reverse transcription insulated isothermal PCR (RT-iiPCR) assay with a portable Pockit nucleic acid analyzer (Tsai et al., 2019). This tool is similar to RT-LAMP (loop-mediated isothermal amplification) assays with the touted advantage of convenience and user-friendliness. The described RT-iiPCR molecular biosensor resolved many of the shortcomings of the first report in 2018 by Tsai et al., where only DENV4 was successfully detected but other arboviruses could also be detected (Refer to Appendix 5 for the latest biosensor's protocol). Lee et al., 2021, successfully developed a CRISPR/Cpf1-based electrochemical biosensor system for DENV-4 RNA detection without the need for amplification (Refer to Appendix 6 for protocol).

Lab-on-chip (LoC) technology allows the real-time detection of dengue by integrating multiple laboratory processes on a microprocessor chip (or a microfluidic chip), which is able to handle minute volumes and hence does not require a large volume of samples from patients. Microfluidics has been widely used in the development of point-of-care dengue diagnostics since the

2000s. A recent development to miniaturize ELISA is using the immunomagnetic agglutination assay (IMA) (Alejo-Cancho et al., 2020). This and other microfluidics designs for dengue diagnostics incorporate a multifunctional micro-transport unit to transfer and mix reagents, thus bringing down not only sample and reagent volumes but also reaction time (Weng et al., 2011). Hosseini et al., 2015, incorporated microsphere technology into their hybrid microfluidics platform based on sandwich ELISA. Aeinehvand et al., 2015, on the other hand, developed centrifugal microfluidic platforms employing microballoon mixers to further reduce assay detection time. Iswardy et al., 2017, developed a bead-based immunofluorescence assay dengue detection using a microfluidic dielectrophoresis platform.

Although the miniaturization of assays in microfluidics helps to reduce cost, paper as the main dengue detection assay platform provides the economic value and simplicity of manufacturing and use (Sekaran & Soe, 2017), as the paper is usable in all existing forms of biorecognition interactions (Table 3.3). To further enhance paper-based diagnostics, the incorporation of paper into microfluidics techniques would provide the next generation of dengue diagnostic assays.

Table 3.3 Classification of paper-based point-of-care tests

Biosensor component	Type
Analyte type	Electronic tongue
	Electronic nose
Options of layout	Lateral flow structure
	Distance-based method
	Microfluidic assay
Dimension of device	2-dimensional
	3-dimensional
Bioreceptor	Antibody-antigen
	Synthetic protein
	Aptamer
	Enzyme
	Tissue
	Microorganism

Biosensor component	Type
Immobilization state	Surface adsorption
	Entrapment
	Cross-linkage
	Covalent bonding
Detection element	Redox indicator
	pH-sensitive indicator
	Inorganic complex
	Nanoparticle (magnetic, carbon, etc.)
	Bimetallic compound
	Nanocluster
Transducer type	Electrochemical
	Optical
	Thermal

Case study

Chong et al., 2021, recently evaluated a microfluidics-based rapid diagnostic test – ViroTrack Dengue Serostate – a semi-quantitative biosensors-based dengue IgG RDT manufactured by BluSense Diagnostics, Copenhagen, Denmark in a Malaysian setting, specifically in a highly urban environment, with capillary blood used as the analyte. The ViroTrack Dengue Serostate is a biosensor consisting of a polymer centrifugal microfluidic cartridge embedded with dry reagent and a portable opto-magnetic reader called BluBox. This biosensor was evaluated with the application of diagnosing recent and previous dengue infections to improve dengue surveillance via sero-epidemiological survey toward strategic and targeted dengue endemic hotspots and potential vaccination strategies. The study by Chong et al., 2021, concluded that ViroTrack Dengue Serostate had sensitivity and specificity of above 90% for the diagnosis of previous and also recent dengue infections when compared to the reference standards, thus indicating good performance and potential field application of this biosensor, albeit with further studies in different populations.

3.2 Biosensor Platform

3.2.1 NS1 Antigen/Antibody Detection

NS1 antigen detection tests detect the non-structural protein NS1 of the dengue virus and have been developed in serum. This protein is secreted into the blood during dengue infection.

Recently, commercially available kits have been developed for the rapid detection of dengue infections during their peak phase. These kits are designed based on the principle of detecting the presence of NS1 antigens and/or anti-dengue antibodies in the blood of suspected dengue patients. NS1 tests are as sensitive as molecular tests during the first 0–7 days of symptoms. After day 7, NS1 tests are not recommended. However, Wang and Sekaran, 2010, mentioned that NS1 is a highly conserved glycoprotein that is present at high concentrations in sera of dengue-infected patients during the early clinical phase of the disease, and it is found from day 1 to day 9 after the onset of fever in a sample of primary or secondary dengue-infected patients.

Various biosensor platforms have been developed for the detection of NSI such as piezoelectric biosensors, optical and electrochemical platforms. Su et al (2003) used an immunochip for identifying DENV NS1 and envelope proteins which was shown to have a 100-fold greater sensitivity than a typical sandwich ELISA technique, but the assay took close to an hour to complete.

Another study (Wu et al., 2005) used a cibacron blue 3 GA gel–heat denature (CB–HD) technique for sample pretreatment procedure and had detection limits of 0.740 g/mL for nonstructural protein 1 (NS1) and 1.727 g/mL for envelope protein (E protein). Another system utilized quartz crystal microbalance (QCM) with molecularly imprinted polymers (MIP) selective for the NS-1 protein of dengue virus serotypes and was able to identify all four DENV serotypes. The MIP-QCM had good sensitivity, a quick operating time and was simple to interpret when detecting the dengue virus. Another benefit of this test was that it did not need serum pre-treatment (Tai et al., 2006; Wang et al., 2018). A localized surface plasmon resonance (LSPR) biosensor was developed (Austin Suthanthiraraj and Sen, 2019) that combines membrane-based blood plasma segregation with a lab-on-chip technology to quickly identify dengue NS1 antigen from a 10 μL volume of whole blood in under 30 min. Cyclic voltammetry (CV) and differential pulse voltammetry (DPV) are two electroanalytical voltammetric techniques that was recently used to detect DENV NS1 antigen (Kim et al., 2019). Another system used an amphoteric-based immunosensor and was able to detect 12 ng/mL with a sensitivity of 85.59 mA/ mMcm^2 (Dias et al., 2013). Using poly (allylamine) and carboxylated carbon nanotubes, another amperometric-based immunosensor electrode was designed and had a limit of detection (LOD) of 0.035 μg/mL (Silva et al., 2015). Apart from these, an electrochemical strategy to identify the dengue NS1 protein biomarker in serum samples and PBS were also to be effective in detecting dengue NS1 (Cecchetto et al., 2017).

3.2.2 Detection of Anti-Dengue IgM Antibodies

Antibodies (Abs), which are amongst the most exquisitely designed and engineered molecules in nature, play a vital role in a number of sensor devices, due to their exquisite target specificity and affinity. Antibodies are ideal biorecognition elements that provide sensors with high specificity and sensitivity (Sharma et al., 2016). This review describes monoclonal and recombinant antibodies and different immobilization approaches crucial for antibody utilization in biosensors (Sharma et al., 2016). A compact, cost-effective, label-free, real-time biosensor based on long-range surface plasmon polariton (LRSPP) gold (Au) waveguides for the detection of dengue-specific immunoglobulin M (IgM) antibody was developed by Jahanshahi et al. (2017). The assay demonstrated detection in actual patient blood plasma samples. They proposed and demonstrated two surface functionalization approaches: (i) a dengue virus serotype 2 (DENV-2) functionalized surface to capture dengue-specific IgM antibody in blood plasma, and (ii) the reverse, a blood plasma functionalized surface to capture DENV-2. The results obtained via these two surface functionalization approaches are comparable to, or of greater quality, than those collected by conventional IgM antibody capture enzyme-linked immunosorbent assay (MAC-ELISA). The second functionalization approach was found to minimize non-specific binding, thus improving the sensitivity and accuracy of the test. They also demonstrate the reuse of biosensors by regenerating the sensing surface down to the virus (or antibody) level or down to the bare Au.

3.2.3 Detection of Dengue Viral Genome

The nucleic acid amplification test (NAAT) is used to detect viral genomic material. It is a confirmatory diagnosis in the detection of dengue virus RNA. The test will be carried out on symptomatic patients during or after 7 days of being detected. Serum, plasma, whole blood, and cerebrospinal fluid are used in molecular tests. There are also a number of NAAT-based platforms in the pipeline that have the potential to be used for the diagnosis of acute DENV infection, e.g., Truelab Real-Time micro-PCR System (Molbio Diagnostics), Gene Xpert Omni Platform (Cepheid), the Alere-i Platform (Alere Inc), the cobas® Liat Analyser (Roche Molecular Diagnostics), and US Centre for Disease Control and Prevention (CDC) DENV 1–4 multiplex assay.

On the other hand, some researchers focused on the viral genome as a target analyte in the hunt for a point-of-care test. Chan, Choong, Perera, & Lim, 2018, developed a genosensor using a silver nanocluster (Ag NC) as a hybridization indicator of target DNA with probe DNA. In the assay, the target DNA was synthesized from dengue RNA extracted from a mosquito and then hybridized with dengue probe DNA and amplified using isothermal amplification. Following the amplification, the product was mixed with $AgNO_3$. Subsequently, the reduction agent $NABH_4$ was added to the mixture to produce Ag NC, which was observed using a UV light. This study showed that the developed genosensor could detect the DENV genome with a LOD of 100 nm. Jin et al. developed an electrochemical genosensor by modifying the graphene oxide surface with SiO_2 particles for enhancing the electrochemical properties of the surface. The negatively charged oligonucleotide primer specific for dengue complementary DNA was immobilized on the graphene-oxide-modified surface. Afterward, the sample containing the target complementary DNA was pipetted. It was found that the oligonucleotide primer hybridized with complementary DNA indicating the presence of dengue DNA in the sample. EIS was carried out on three-electrode cells using PBS solution, and impedance spectra were measured. Dengue DNA was successfully detected with a LOD of 1 femtomolar concentration of the viral genome. Surface plasmon resonance (SPR) is a medical diagnostic technique with high sensitivity and specificity. Jahanshahi (2015) developed a new method based on SPR as a rapid, 10-min detection of the anti-dengue virus in human serum samples (Jahanshahi, Zalnezhad, Sekaran, & Adikan, 2015). This novel technique, known as rapid immunoglobulin M (IgM)-based dengue diagnostic test, can be utilized quickly and easily at point-of-care facilities. Four dengue virus serotypes were used as ligands on a biochip. According to the results, a serum volume of only 1 ml from a dengue patient (as a minimized volume) is required to indicate SPR angle variation to determine the ratio of each dengue serotype in samples with 83–93% sensitivity and 100% specificity. Using mass spectrometry, biosensor technology using mass spectrometry can provide rapid discrimination of biological components in complex mixtures by producing a specific molecular profile of the bacteria or virus analyzed. These organisms can then be identified and quantified using inbuilt software, which enables comparisons between the mass spectra and databases of infectious agents. This could be useful in determining both dengue serotypes and genotypes during an outbreak. Kits for infectious agents are available and can be designed to meet specific requirements. Samples are processed for DNA extraction, PCR amplification, mass spectrometry, and computer analysis.

3.3 Limitations

The ability of biosensors to rapidly and specifically detect a wide range of molecules makes them highly relevant to medical and scientific applications. Approaches to biosensor design are as diverse as their applications, with major biosensor classes including nucleic acids, proteins, and transcription factors. Each type has advantages and limitations based on the intended application and the parameters that are required for optimal performance. Specifically, the choice of biosensor design must consider factors such as the ligand specificity, sensitivity, dynamic range, functional range, mode of output, time of activation, ease of use, and ease of engineering. Biosensors, like all other detection systems, also exhibit saturation limits, as a result of the materials used in designing the sensor. It also has a limit of detection, which is the lowest amount of substance that a biosensor can quantify. This is important for applications in the early detection of the disease. Within those first few days, you may be under the limit of detection so the biosensor may produce a negative result. False positives can also occur and these may also be a result of human error. At any rate, the probability of false positives or false negatives should be close to zero for an effective biosensor or diagnostic test, but we need to understand that these things can happen, and the reasons why they can happen for a given application. Other possible reasons include insufficient sample collection, poor sample detection, and even degradation of the sensing components over time, ultimately rendering the biosensor useless. These are some of the issues faced when designing a new biosensor. Another question to ask is what amount of accuracy is actually necessary? This usually depends on the severity of the application. Now, true accuracy and predictive accuracy are also measures of biosensor performance. Artificial intelligence and machine learning have their place in the future of biosensors and health analysis and research and development by many investigators are looking at how to overcome these flaws, so we can have more effective applications that address much-needed problems in healthcare.

References

Aeinehvand, M. M., Ibrahim, F., Harun, S. W., Djordjevic, I., Hosseini, S., Rothan, H. A., ... Madou, M. J. (2015). Biosensing enhancement of dengue virus using microballoon mixers on centrifugal microfluidic platforms. *Biosensors and Bioelectronics, 67*, 424–430. https://doi.org/10.1016/j.bios.2014.08.076

Alejo-Cancho, I., Navero-Castillejos, J., Peiró-Mestres, A., Albarracín, R., Barrachina, J., Navarro, A., ... Martínez, M. J. (2020). Evaluation of a novel microfluidic immuno-magnetic agglutination assay method for detection of dengue virus NS1 antigen. *PLOS Neglected Tropical Diseases, 14*(2), e0008082. https://doi.org/10.1371/journal.pntd.0008082

Austin Suthanthiraraj, P. P., & Sen, A. K. (2019). Localized surface plasmon resonance (LSPR) biosensor based on thermally annealed silver nanostructures with on-chip blood-plasma separation for the detection of dengue non-structural protein NS1 antigen. *Biosens. Bioelectron. 132*, 38–46. https://doi.org/10.1016/j.bios.2019.02.036

Bazin, I., Tria, S. A., Hayat, A., & Marty, J.-L. (2017). New biorecognition molecules in biosensors for the detection of toxins. *Biosensors and Bioelectronics, 87,* 285–298. https://doi.org/10.1016/j.bios.2016.06.083

Bhalla, N., Jolly, P., Formisano, N., & Estrela, P. (2016). Introduction to biosensors. *Essays in Biochemistry, 60*(1), 1–8. https://doi.org/10.1042/EBC20150001

Cecchetto, J., Fernandes, F. C. B., Lopes, R., Bueno, P. R. (2017). The capacitive sensing of NS1 Flavivirus biomarker. *Biosens. Bioelectron. 87*, 949–956. https://doi.org/10.1016/j.bios.2016.08.097"10.1016/j.bios.2016.08.097

Chan, S. K., Choong, Y. S., Perera, D., & Lim, T. S. (2018). Dengue serotyping with a label-free DNA sensor. *Analytical Methods, 10*(2), 214–222. https://doi.org/10.1039/C7AY02131C

Chong, Z. L., Soe, H. J., Ismail, A. A., Mahboob, T., Chandramathi, S., & Sekaran, S. D. (2021). Evaluation of the diagnostic accuracy of a new biosensors-based rapid diagnostic test for the point-of-care diagnosis of previous and recent dengue infections in Malaysia. *Biosensors, 11*(5), 129. https://doi.org/10.3390/bios11050129

den Boef, G. & Hulanicki, A. (1983). Recommendations for the usage of selective, selectivity and related terms in analytical chemistry. *Pure and Applied Chemistry, 55*(3), 553–556. https://doi.org/10.1351/pac198855030553

Dias, A. C. M. S., Gomes-Filho, S. L. R., Silva, M. M. S., & Dutra, R. F. (2013). A sensor tip based on carbon nanotube-ink printed electrode for the dengue virus NS1 protein. *Biosens. Bioelectron. 44*, 216–221. https://doi.org/10.1016/j.bios.2012.12.033

Faria, H. A. M. & Zucolotto, V. (2019). Label-free electrochemical DNA biosensor for zika virus identification. *Biosensors and Bioelectronics, 131*, 149–155. https://doi.org/10.1016/j.bios.2019.02.018

Goode, J. A., Rushworth, J. V. H., & Millner, P. A. (2015). Biosensor regeneration: A review of common techniques and outcomes. *Langmuir, 31*(23), 6267–6276. https://doi.org/10.1021/la503533g

Hosseini, S., Aeinehvand, M. M., Uddin, S. M., Benzina, A., Rothan, H. A., Yusof, R., ... Ibrahim, F. (2015). Microsphere integrated microfluidic disk: Synergy of two techniques for rapid and ultrasensitive dengue detection. *Scientific Reports, 5*(1), 16485. https://doi.org/10.1038/srep16485

Iswardy, E., Tsai, T.-C., Cheng, I.-F., Ho, T.-C., Perng, G. C., & Chang, H.-C. (2017). A bead-based immunofluorescence-assay on a microfluidic dielectrophoresis platform for rapid dengue virus detection. *Biosensors and Bioelectronics, 95*, 174–180. https://doi.org/10.1016/j.bios.2017.04.011

Jahanshahi, P., Wei, Q., Jie, Z., Ghomeishi, M., Sekaran, S. D., & Mahamd Adikan, F. R. (2017). Kinetic analysis of IgM monoclonal antibodies for determination of dengue sample concentration using SPR technique. *Bioengineered, 8*(3), 239–247. https://doi.org/10.1080/21655979.2016.1223413

Jahanshahi, P., Zalnezhad, E., Sekaran, S. D., & Adikan, F. R. M. (2015). Rapid immunoglobulin M-based dengue diagnostic test using surface plasmon resonance biosensor. *Scientific Reports, 4*(1), 3851. https://doi.org/10.1038/srep03851]

Kim, J. H., Cho, C. H., Ryu, M. Y., Kim, J.-G., Lee, S.-J., Park, T. J., & Park, J. P. (2019). Development of peptide biosensor for the detection of dengue fever biomarker, nonstructural 1. *PLoS One 14*, e0222144. https://doi.org/10.1371/journal.pone.0222144

Lee, Y., Choi, J., Han, H.-K., Park, S., Park, S. Y., Park, C., ... Min, J. (2021). Fabrication of ultrasensitive electrochemical biosensor for dengue fever viral RNA based on CRISPR/Cpf1 reaction. *Sensors and Actuators B: Chemical, 326*, 128677. https://doi.org/10.1016/j.snb.2020.128677

Lisdat, F. & Schäfer, D. (2008). The use of electrochemical impedance spectroscopy for biosensing. *Analytical and Bioanalytical Chemistry, 391*(5), 1555–1567. https://doi.org/10.1007/s00216-008-1970-7

Long, G. L. & Winefordner, J. D. (1983). Limit of detection a closer look at the IUPAC definition. *Analytical Chemistry, 55*(07), 712A-724A. https://doi.org/10.1021/ac00258a724

Miserere, S. & Merkoçi, A. (2015). Microfluidic electrochemical biosensors: Fabrication and applications. *Lab-on-a-Chip Devices and Micro-Total Analysis Systems: A Practical Guide.* J. Castillo-León & W. E. Svendsen (Eds.). pp. 141–160. Cham: Springer International Publishing. https://doi.org/10.1007/978-3-319-08687-3_6

Persson, B. (2001). The use of selectivity in analytical chemistry: Some considerations. *TrAC Trends in Analytical Chemistry, 20*(10), 526–532. https://doi.org/10.1016/S0165-9936(01)00093-0

Sekaran, S. D. & Soe, H. J. (2017). Issues in contemporary and potential future molecular diagnostics for dengue. *Expert Review of Molecular Diagnostics, 17*(3), 217–223. https://doi.org/10.1080/14737159.2017.1275963

Shamala, D. S. (2015). Laboratory diagnosis of dengue: A review. *International Medical Journal Malaysia, 14*(1), 17–28. https://doi.org/10.31436/imjm.v14i1.452

Sharma, S., Byrne, H., & O'Kennedy, R. J. (2016). Antibodies and antibody-derived analytical biosensors. *Essays in Biochemistry, 60*(1), 9–18. https://doi.org/10.1042/EBC20150002

Shrivastava, A. & Gupta, V. B. (2011). Methods for the determination of limit of detection and limit of quantitation of the analytical methods. *Chronicles of Young Scientists, 2*, 21–25.

Silva, M. M. S., Dias, A. C. M. S., Silva, B. V. M., Gomes-Filho, S. L. R., Kubota, L. T., Goulart, M. O. F., & Dutra, R. F. (2015). Electrochemical detection of dengue virus NS1 protein with a poly(allylamine)/carbon nanotube layered immunoelectrode. *J. Chem. Technol. Biotechnol. 90*, 194–200. https://doi.org/10.1002/jctb.4305

Su, C.-C., Wu, T.-Z., Chen, L.-K., Yang, H.-H., & Tai, D.-F. (2003). Development of immunochips for the detection of dengue viral antigens. *Anal. Chim. Acta 479*, 117–123. https://doi.org/10.1016/S0003-2670(02)01529-5

Tai, D.-F., Lin, C.-Y., Wu, T.-Z., Huang, J.-H., & Shu, P.-Y. (2006). Artificial receptors in serologic tests for the early diagnosis of dengue virus infection. *Clin. Chem. 52*, 1486–1491. https://doi.org/10.1373/clinchem.2005.064501

Tsai, J.-J., Liu, W.-L., Lin, P.-C., Huang, B.-Y., Tsai, C.-Y., Lee, P.-Y. A., ... Chen, C.-H. (2019). A fully automated sample-to-answer PCR system for easy and sensitive detection of dengue virus in human serum and mosquitos. *PLOS ONE, 14*(7), e0218139. https://doi.org/10.1371/journal.pone.0218139

Wang, H., Xu, Q., Wang, J., Du, W., Liu, F., & Hu, X. (2018). Dendrimer-like amino-functionalized hierarchical porous silica nanoparticle: a host material for 2,4-dichlorophenoxyacetic acid imprinting and sensing. *Biosens. Bioelectron. 100*, 105–114. https://doi.org/10.1016/j.bios.2017.08.063

Weng, C.-H., Huang, T.-B., Huang, C.-C., Yeh, C.-S., Lei, H.-Y., & Lee, G.-B. (2011). A suction-type microfluidic immunosensing chip for rapid detection of the dengue virus. *Biomedical Microdevices, 13*(3), 585–595. https://doi.org/10.1007/s10544-011-9529-3

Wu, T.-Z., Su, C.-C., Chen, L.-K., Yang, H.-H., Tai, D.-F., & Peng, K.-C. (2005). Piezoelectric immunochip for the detection of dengue fever in viremia phase. *Biosens. Bioelectron. 21*, 689–695. https://doi.org/10.1016/j.bios.2004.12.019

Chapter 4

Search for Newer Markers for Dengue Disease

Yong Yean Kong,[a] Tan Hong Yen,[b] Shamala Devi Sekaran,[c] and Rishya Manikam[d]

[a]*School of Traditional Chinese Medicine, Xiamen University Malaysia, Sepang, Selangor, Malaysia*

[b]*Laboratory, Xiamen University Malaysia, Sepang, Selangor, Malaysia*

[c]*Faculty of Applied Sciences, UCSI University Kuala Lumpur Campus, Malaysia*

[d]*Emergency and Trauma Centre, University Malaya Medical Centre, Kuala Lumpur, Malaysia*

shamala@ucsiuniversity.edu.my, yong.yeankong@gmail.com

4.1 Introduction

Dengue disease is the most important arboviral infection across the tropical and subtropical regions affecting ~400 million people each year, of which 96 million manifest as symptomatic infection (Messina, Brady et al., 2014). The disease has a wide range of clinical manifestations ranging from mild febrile illness to more severe ones

Dengue Diagnostics: The Right Test at the Right Time for the Right Group
Edited by Shamala Devi Sekaran

ISBN 978-981-4968-97-3 (Hardcover), 978-1-032-66956-4 (eBook)
www.jennystanford.com

including bleeding, endothelial dysfunction, capillary leakage, and multiple organ failure seen in severe dengue hemorrhagic fever (DHF) or dengue shock syndrome (DSS) (Rathakrishnan, Klekamp et al., 2014). While most dengue patients have mild/moderate illness, about 5% of hospitalized dengue patients could develop severe dengue (WHO, 2009). These severe manifestations usually occur late, typically on days 4–7 (during the defervescence phase) after the onset of illness; hence, potentially there is a window of opportunity that allows the identification of patients who are more likely to progress. During a dengue endemic, thousands of patients are seen at the outpatient clinic daily, bringing a great strain to the healthcare system (Paranavitane, Gomes et al., 2014). The ability to predict which patients will develop life-threatening dengue disease will be a great advantage in terms of patient management as well as prioritizing medical resources for intervention (Carrasco, Leo et al., 2014). Though the WHO provided some simple clinical and biochemical parameters as "warning signs" for severe dengue in the 2009 Dengue Guideline for Diagnosis, Treatment, Prevention, and Control (WHO, 2009), one cannot predict which patients will progress to severe dengue based on those parameters. Therefore, there is a pressing need to identify biomarkers that prognose severe dengue. Biomarkers here are defined as measurable parameters (physical, chemical, or biological) that can indicate a pathogenic process or progression of dengue disease. There have been many studies on the biomarkers that could be associated with the severity of the disease, especially in the last decade. This chapter will summarize major findings in the field of biomarkers that predict/prognose the clinical outcomes of dengue disease.

4.2 Viral Markers and Their Association with Dengue Severity

In any infectious disease, the load of the pathogen is often associated with the severity of the disease. A high plasma level of dengue viral load and higher amount of dengue-infected cells have been linked to the severity of dengue disease, where severe dengue usually had

10–100 fold higher dengue viral load than the DF (Vaughn, Green et al., 2000; Libraty, Endy et al., 2002). The viral load is usually quantified using RT-PCR assay (Houng, Chung-Ming Chen et al., 2001; Kong, Thay et al., 2006; Yong, Thayan et al., 2007; Waggoner, Abeynayake et al., 2013) (Figure 4.1). Vaughn et al. showed that the peak of the viral load correlated with the severity of dengue disease in the subsequent days (Vaughn, Green et al., 2000); therefore, the peak of the viral load may be a useful biomarker for the development of severe dengue disease. However, such approaches require longitudinal sampling from the early phase of the disease onset, hence operationally difficult and may not be applicable to all cases, especially those who are diagnosed late.

The non-structural protein 1 (NS1) antigen is a glycoprotein that is pivotal for dengue replication and is secreted by the infected cell to the circulation. The levels of NS1 antigenemia within 72 hours of fever onset have been shown to correlate with the level of dengue viremia and disease severity (Libraty, Young et al., 2002). Given that NS1 is also shown to activate the complement system and hence may play a role in increasing permeability during severe dengue, it is an ideal biomarker for both diagnosis and prognosis. Several commercial assays for NS1 detection are currently used in clinical settings and generally, can be divided into two formats, i.e., rapid diagnostic test (RDT) and ELISA. But, since quantitation of NS1 level is required for prognosis purposes, the ELISA format must be used. The three NS1 ELISA kits currently commercially available in the market are: (i) Platelia Dengue NS1 Ag ELISA (Bio-Rad, France), (ii) DENV Detect NS1 ELISA (InBios International, U.S.), and (iii) Panbio Dengue Early ELISA (2nd generation) (Alere, U.S.) (Section 4.2). More recently, by using multivariate analysis, Huang et al. showed that NS1 was an independent risk of more severe dengue disease (Huang, Tsai et al., 2020). Nonetheless, several other factors, such as host immunity, day of fever onset, and viral serotype (DENV 2 has significantly lower NS1 than DENV1) have been shown to affect the levels of NS1 in serum, hence affecting the overall sensitivity of NS1 as a potential prognosis marker (Duong, Ly et al., 2011; Duyen, Ngoc et al., 2011).

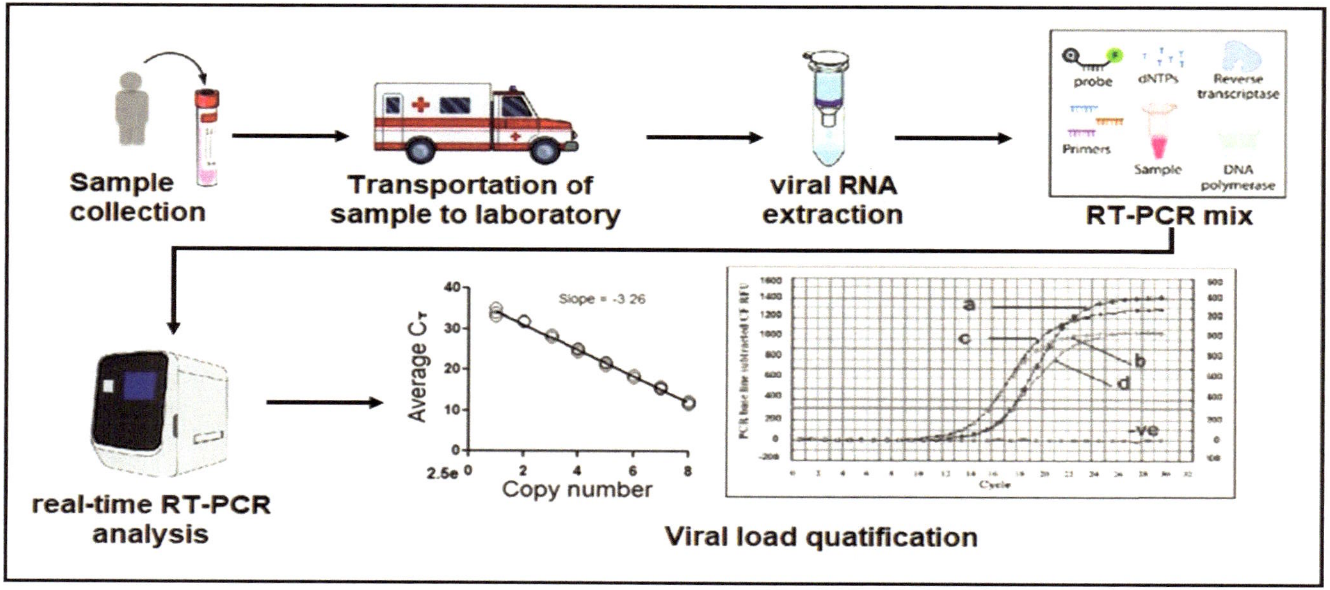

Figure 4.1 Typical steps for the detection and quantification of dengue virus with real-time RT-PCR.

4.3 Clinical Markers of Severe Dengue

Clinical symptoms: Dengue infections cause a wide range of clinical manifestations, from asymptomatic to mild, self-limited fever typically accompanied by muscle and joint pain, rashes, nausea, vomiting, and diarrhea. Some patients may recover days after while others may deteriorate and develop warning signs of tenderness and abdominal pain, vomiting, bleeding, fluid accumulation, hepatomegaly, increase of hematocrit, as well as thrombocytopenia (WHO, 2009; Rathakrishnan, Klekamp et al., 2014). In this critical phase, if proper medical support is not given in time, the patients may further progress into severe dengue characterized by severe plasma leakage, bleeding, organs dysfunction, hypovolemic shock, and death (WHO, 2009; Simmons, Farrar et al., 2012).

Since patients with mild and warning signs can develop into severe dengue disease later [2], it is important to look for signs and symptoms to facilitate the prognosis of the progression into severe dengue disease. Besides, the clinical manifestation is often inexpensive and hence it is often used as a tool for monitoring disease progression (Pawitan, 2011; Yacoub, Mongkolsapaya et al., 2013). Vomiting, myalgia, arthralgia, diarrhea, headaches, retro-orbital pain, nausea, rashes, and petechia are common among dengue patients; but symptoms like concurrent increase in hematocrit with the decrease of platelets, abdominal pain, bleeding tendencies, postural giddiness, hepatomegaly and/or splenomegaly, pleural effusion, ascites as well as acute hepatitis are indications for ICU admission (Rathakrishnan, Klekamp et al., 2014). Among these all, bleeding (hematemesis and melena), abdominal pain, skin rashes, and hepatomegaly/acute hepatitis are the most prominent symptoms associated with severe dengue disease (Leo, Gan et al., 2013; Zhang, Zhou et al., 2014; Chen, Chan et al., 2016; Fernando, Wijewickrama et al., 2016; Padyana, Karanth et al., 2019) (Figure 4.2). Furthermore, the elevation of liver enzymes, i.e., alanine aminotransferase (ALT) and aspartate aminotransferase (AST) levels >1000 U/l have also been identified as predictors for organ failure and death (Lee, Gan et al., 2012; Kittitrakul, Silachamroon et al., 2015; Fernando, Wijewickrama et al., 2016). Nonetheless, one study reported that most of these warning signs occur within one day before progressing to severe illness. The narrow window allowing for effective intervention can be challenging especially in resource-limited settings (Leo, Gan et al., 2013), and hence laboratory markers are needed for early prognosis.

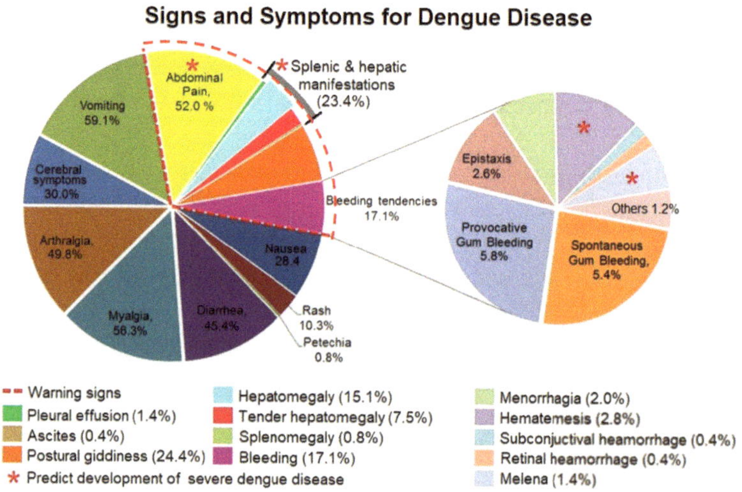

Figure 4.2 Clinical symptoms commonly reported with dengue disease and warning signs that predict severe dengue disease (Rathakrishnan, Klekamp et al., 2014). Red-dotted line (– –) represents warning signs that are often reported with severe dengue. Red (*) represents warning signs that are predicted severe dengue disease (Zhang, Zhou et al., 2014).

4.4 Markers of Immune Activation and Severe Dengue Disease

In dengue disease, the severe clinical manifestation usually occurs late in the illness when the peak of dengue viremia had passed (Simmons, Farrar et al., 2012). Furthermore, severe dengue disease is more commonly found in secondary infection with a different dengue serotype (heterotypic infection) (Zompi and Harris, 2013), indicating that the pathogenesis of dengue disease is at least in part immune-mediated (Lei, Yeh et al., 2001). A phenomenon known as antibody-dependent enhancement (ADE) is particularly implicated in the pathogenesis of severe dengue (Kontny, Kurane et al., 1988; Littaua,

Kurane et al., 1990; Rodrigo, Jin et al., 2006; Moi, Lim et al., 2010), where during a heterotypic infection, the memory B cells formed by previous infection are predominantly expanded, producing a large number of antibodies targeting to previous dengue serotype. Such antibodies are usually non-neutralizing yet form immune complexes that will later be taken up by Fcγ receptor-bearing cells (e.g., B cells, monocytes, macrophages, dendritic cells, and mast cells), resulting in an increased dengue-infected cell mass, and hence higher viral load and more severe dengue diseases (Littaua, Kurane et al., 1990). The dengue-infected immune cells are also hyperactivated, leading to the release of a massive amount of cytokine (known as cytokine storm) (Rothman, 2011; Srikiatkhachorn, Mathew et al., 2017; Kuczera, Assolini et al., 2018) and other vasoactive mediators (causing endothelial dysfunction) (Yacoub, Lam et al., 2016), and hence more severe disease manifestation. As cytokines and vasoactive mediators are soluble biomarkers, they are usually measured either singly by ELISA or as a multiplex by bead suspension array (Figure 4.3).

Other immune activation markers associated with dengue severity include the magnitude of T cells, NK cell responses, and inflammasome activation in monocyte/macrophage. While CD4 and CD8 T cells played a pivotal role in controlling a viral infection (de Matos, Carvalho et al., 2015), aberrant activation of CD4 and CD8 T cell led to more severe dengue disease (Green, Pichyangkul et al., 1999; Green, Vaughn et al., 1999). The aberrant activation of T cells during dengue infection could be due to direct infection of these T cells by the dengue virus via interaction with heparin sulfate expressed on the T cell surface. The infected T cell becomes activated and expresses CD69, HLA-DR, and CD38, and secrets pro-inflammatory mediators (Silveira, Wowk et al., 2018). While the HLA-DR and CD38 expressing CD4 and CD8 T cells are expanded during acute infection, only a small fraction of these cells was actually specific to dengue antigen (Figure 4.4, B2). The remaining HLA-DR and CD38-expressing T cells are bystander cells that do not contribute to the viral control (Chandele, Sewatanon et al., 2016).

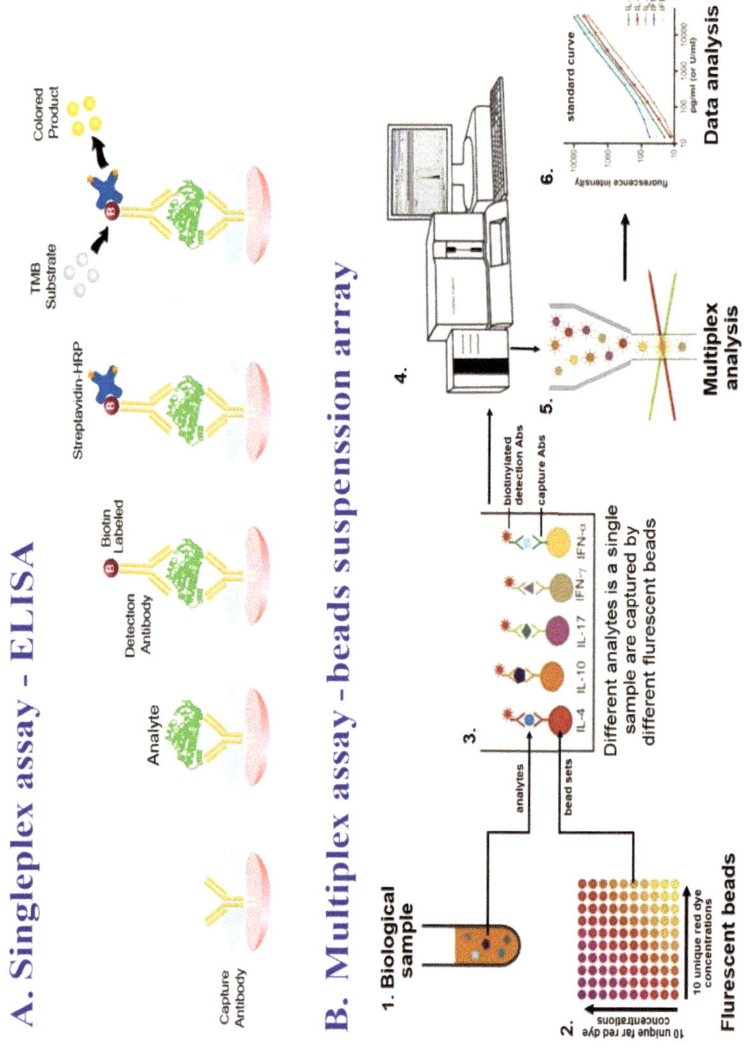
A. Singleplex assay - ELISA
Capture Antibody
Analyte
Detection Antibody
Biotin Labeled
Streptavidin-HRP
TMB Substrate
Colored Product
B. Multiplex assay -beads suspenssion array
1. Biological sample
2.
10 unique far red dye concentrations
10 unique red dye concentrations
Flurescent beads
analytes
bead sets
3.
IL-4
IL-10
IL-17
IFN-γ
IFN-α
biotinylated detection Abs
capture Abs
Different analytes is a single sample are captured by different flurescent beads
4.
5.
Multiplex analysis
6.
standard curve
fluorescence intensity
pg/ml (or U/ml)
Data analysis

Figure 4.3 Singleplex vs. multiples assay. (A) Singleplex assay – ELISA: the method is the same as described above. (B) Multiplex assay – bead suspension array. 1. Biological sample containing multiple analytes (cytokines, vasoactive mediators, inflammatory mediators, and other soluble proteins) is obtained from the patient. 2. Dead suspension array uses microsphere or beads (polystyrene or magnetic), stained with 10 different concentrations of two dyes (i.e., red and infrared) to create 100 distinct bead sets. Each bead set is then conjugated with monoclonal antibodies specific to a target analyte. This system allows multiple analytes to be measured in a single reaction. 3. Once the analyte of interest has bound with the primary capture antibody, a secondary biotinylated detection antibody is then added to the sample mix. 4–6. The sample mix will then be subjected to a bead suspension array reader where the beads will flow through a laser beam that allows the identification of beads (specific analyte of interest) and quantifies the fluorescence level of each bead. By incorporating the serial dilution of known concentration standards for each analyte, the concentration of each analyte of interest in a biological sample can be measured. There are two bead suspension array systems commercially available in the market, i.e., Bioplex system and Luminex system, of which both operate at the same principle.

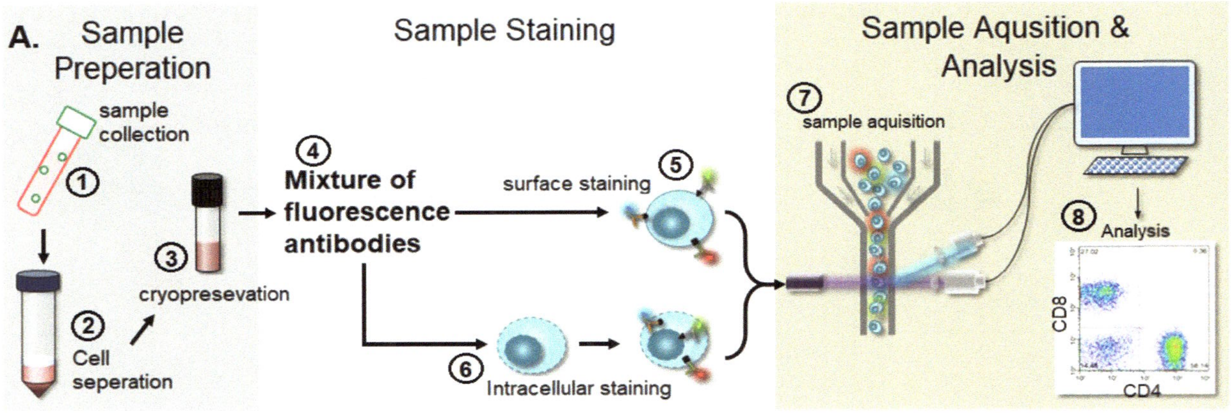

(*Figure contd.*)

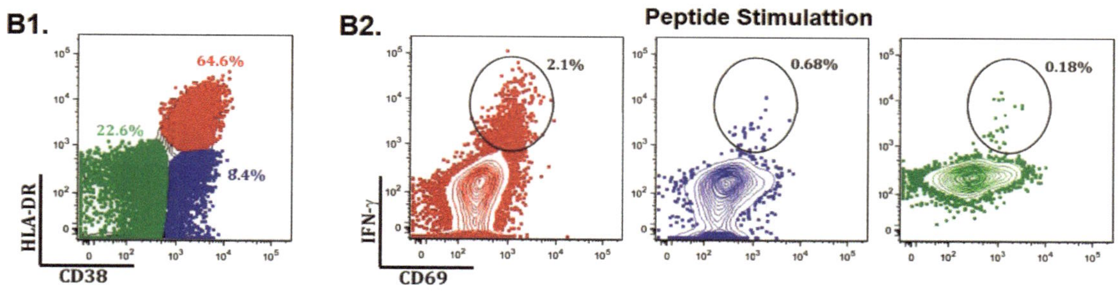

Figure 4.4 Typical process of flow cytometry experiment. A. 1. Peripheral blood collected from patients. 2. Sample preparation involves Ficoll gradient separation of peripheral blood mononuclear cells (PBMC). 3. Cryopreservation, until used. 4. PBMC is then stained with fluorescently conjugated antibodies. 5. Surface staining is used for immunophenotyping. 6. Intracellular staining is used to stain cytokines expressed in respective cells. 7–8. The stained cell will then be subjected to a flow cytometer for analysis. B1. Expression of immune activation markers, i.e., HLA-DR and CD38 on CD4 and CD8 T cells. B2. Expression of IFNγ on HLA-DR$^+$CD38$^+$CD69$^+$ subset (red), HLA-DR$^-$CD38$^+$CD69$^+$ subset (blue), and HLA-DR$^-$CD38$^-$CD69$^+$ subset (green) after stimulated with dengue antigens.

Note: Although the dengue patient had ~65% of immune activated T cells (HLA-DR$^+$CD38$^+$), only ~2% among these activated cells in producing IFNγ specific to dengue antigen, indicating that the majority of the activated T cells do not contribute to antiviral responses. Figures 4.4 B1 and B2 are adapted from (Chandele, Sewatanon et al., 2016). A detailed study of T cell response against dengue antigens identified 17 different immunogenic peptides scattered along the dengue genome (Figure 4.5). Among the 17 peptides, patients with DHF have particularly strong responses toward peptides, i.e., $NS3_{422-431}$ and $NS5_{563-571}$ (Figure 4.6). The two peptides may be used to measure the magnitude of anti-dengue-specific immune responses in predicting disease severity.

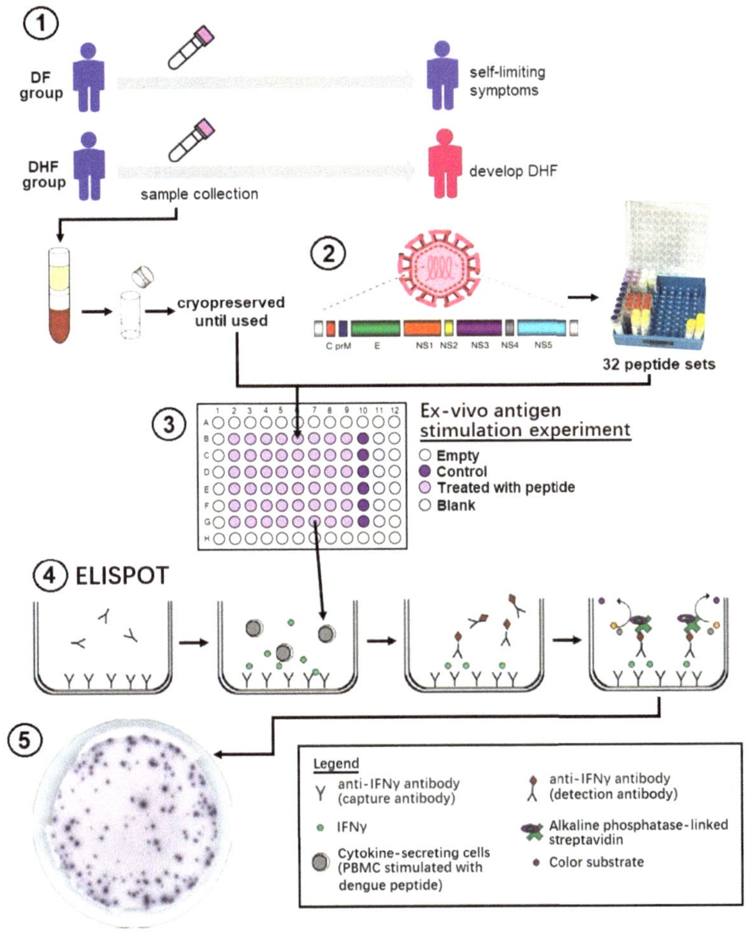

Figure 4.5 Typical workflow for ELISPOT experiment. 1. Sample collection. Peripheral blood is collected from DF and DHF patients. The PBMC is then extracted using the Ficoll gradient separation method and cryopreserved until used. 2. 32 peptide antigens spanning across the entire dengue genome, i.e., C, E, NS2B, NS3, NS4A, NS4B, and NS5, are synthesized. 3. Ex vivo antigen stimulation experiment, where PBMCs from DF and DHF patients are cultured (20 hours) with and without dengue peptides. The dengue antigens will stimulate memory T cell cells to produce IFNγ. 4. The number of cells responding to peptide stimulation is then measured by using an ELISPOT assay, where each spot represents one IFNγ-producing cell. 5. The number of spots will then be measured by an ELISPOT reader.

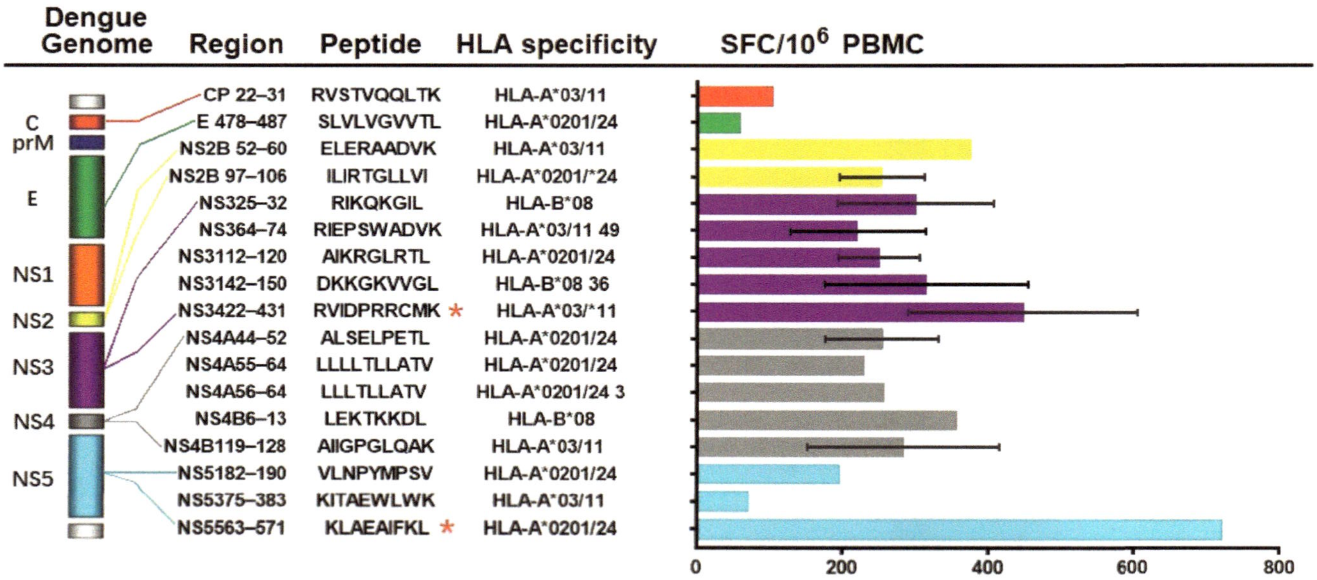

Figure 4.6 T cell responses of DF and DHF patients to individual peptides. SFC, spot forming cell. Red (*) indicates the peptide that preferentially responded to by DHF patients. Figure modified from (Appanna, Huat et al., 2007).

The numbers and immunophenotyping of NK cells in blood also had an association with dengue severity (Green, Pichyangkul et al., 1999; Wahid, Sanusi et al., 2000; Azeredo, De Oliveira-Pinto et al., 2006). Approximately 90% of NK cells in human PBMC are $CD56^{lo}$ and $CD16^{hi}$, expressing transcription factors, $T\text{-}bet^{hi}$ and $Eomes^{lo}$. These phenotypes of NK cells are considered mature and with high cytotoxicity. The remaining 10% are $CD56^{hi}$ and $CD16^{lo}$, expressing transcription factors, $T\text{-}bet^{hi}$ and $Eomes^{hi}$. This type of NK cell is less mature and less cytotoxic as compared to $CD56^{lo}$ NK cells. But they are robust cytokine producers (Luetke-Eversloh, Killig et al., 2013) (Figure 4.7). During dengue infection, the frequency of cytokine-producing NK cells (i.e., $CD56^{hi}CD16^{lo}$ subset) and the frequencies of immune activation markers (i.e., CD69) expressing cytotoxic NK cells (i.e., $CD56^{lo}CD16^{hi}$ subset) are found to be elevated among DF patients compared to SD patients (Gandini, Petitinga-Paiva et al., 2017; Keawvichit, Khowawisetsut et al., 2018). A swift NK cell response during the febrile phase of dengue infection is pivotal for effective viral control, resulting in a reduced inflammatory response and hence reduced disease severity.

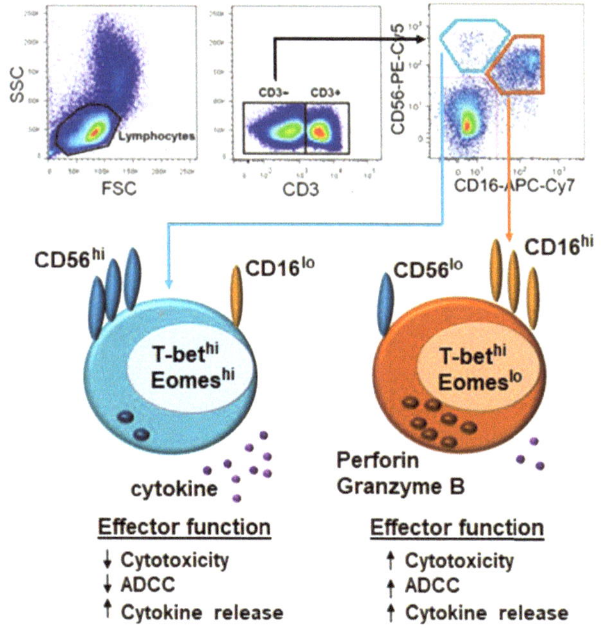

Figure 4.7 NK cells subpopulation.

4.5 Cytokines in Dengue Infection

The use of cytokines as prognostic markers for severe dengue disease has certain advantages. As severe dengue is essentially a systemic inflammation that involves cytokine storm, some of these signature cytokines are elevated early during the course of the disease, allowing for early detection (Puc, Ho et al., 2021). Secondly, they are relatively simple and inexpensive to be measured. With the advent of the bead suspension array, multiple cytokine/inflammatory mediators can be measured in one go, hence allowing for disease profiling and more accurate prediction (Parsa, Vafajoo et al., 2018). The levels of immune activation markers, including cytokines, chemokines, and inflammatory mediators, have been shown to predict outcomes in dengue in some small studies. In general, early responding immunity is pivotal in controlling viral infection. Studies showed that an increase of pro-inflammatory cytokines, such as IFNγ, IFNα, TNFα, and IL-12, at the febrile phase was associated with less severe dengue disease. Other pro-inflammatory cytokines, such as IL-2, IL-6, IL-1β, IL-18, and IL-17, along with other Th2 cytokines (e.g., IL-4, IL-10) and chemokine (e.g., CXCL8, CXCL10, MCP-1) were associated with more severe dengue disease. The results are summarized in Table 4.1.

Cytokine storm is an immune pathogenic condition often experienced by severe dengue patients, where many pro-inflammatory cytokines are mass-produced, causing cellular dysfunction, tissue damage, organ failures, and death (Shrivastava, Valenzuela Leon et al., 2020). IL-1β and IL-18 are pro-inflammatory cytokines produced during inflammatory disease following inflammasome activation and their level had been shown to correlate with the severity of dengue disease (Bozza, Cruz et al., 2008; Jaiyen, Masrinoul et al., 2009; Yong, Tan et al., 2017). As an endo-pyrogen, IL-1β can stimulate lymphocytes infiltration to the site of infection (Dinarello, 2011) while IL-18 is known to facilitate inflammatory responses in various immune cells (e.g., monocyte, macrophage, T cell, NK cells, basophile, and eosinophil) to produce a wide range of pro-inflammatory mediators including cytokines (e.g., IFNγ, TNFα, and IL-6), chemokines (CXCL-10, MCP-1, and MIP-1), and adhesion molecule (ICAM-1) (Nanda, Ho et al., 2021). The M protein of the dengue virus is known to cause activation of the NLRP3 inflammasome (Pan, Zhang et al., 2019) in

Table 4.1 Cytokine, chemokine, and inflammatory mediator associated with severe dengue

	Marker measured at		Associated with SD		Remark	Reference
Marker	**Febrile**	**Defervescence**	**↑ risk**	**↓ risk**		
				Type 1 cytokine		
IFNγ	•			•	–	Chen, Liu et al., 2005; Chen, Yang et al., 2007
	•	•		•	Non-survival of SD has a very low IFNγ response.	Chen, Lei et al., 2006
		•		•	–	Kumar, Liang et al., 2012
		•		•	SD and DHF patients have significantly lower IFNγ responses compared to DF patients.	Soundravally, Hoti et al., 2014
	•		•		–	Malavige, Gomes et al., 2013
	•	•	•		Elevation at the defervescence phase associated with SD.	Bozza, Cruz et al., 2008; Butthep, Chunhakan et al., 2012
	•	•		•	Serial samples were collected from day 1 to day 12. The result showed IFNγ peak at days 4–6 (febrile phase) and the peak level was associated with ↓ severity.	Liao, Tang et al., 2015

(Continued)

Table 4.1 (*Continued*)

	Marker measured at		Associated with SD		Remark	Reference
Marker	**Febrile**	**Defervescence**	**↑ risk**	**↓ risk**		
	•	•	•		Serial samples were collected from day 1 to day 18. The result showed that the high IFNγ defervescence phase was associated with ↑ severity.	Zhao, Huang et al., 2016
IFNα	•			•	–	Chen, Yang et al., 2007; Michels, de Mast et al., 2015; Singla, Kar et al., 2016
TNFα	•			•	The higher expression of membrane-bound TNF receptors (mTNFR1 & mTNFR2) was associated with reduced risk of SD.	Wang, Chen et al., 2007
		•		•	While the level of TNFα was associated with less severe dengue disease; the levels of soluble TNFα receptors (sTNFR1 & sTNFR2) were associated with an increased risk of SD.	Chunhakan, Butthep et al., 2015; Arias, Valero et al., 2014
	•		•		–	Houghton-Trivino, Salgado et al., 2010

	Marker measured at		Associated with SD		Remark	Reference
Marker	Febrile	Defervescence	↑ risk	↓ risk		
		•	•		–	Soundravally, Hoti et al., 2014
IL-2	•		•		–	Senaratne, Carr et al., 2016
IL-12	•			•	A high level of IL-12 was accompanied by increased mTNFR1 and mTNFR2.	Arias, Valero et al., 2014
	•	•		•	The level of IL-12 was significantly low in patients with SD.	Singla, Kar et al., 2016
IL-6	•	•	•		The level of IL-6 was elevated in SD patients and was non-survival.	Chen, Lei et al., 2006; Bozza, Cruz et al., 2008; Butthep, Chunhakan et al., 2012; Guerrero, Arrieta et al., 2013; van de Weg, Pannuti et al., 2013; Arias, Valero et al., 2014; Liao, Tang et al., 2015; Singla, Kar et al., 2016

(*Continued*)

Table 4.1 (*Continued*)

	Marker measured at		Associated with SD		Remark	Reference
Marker	**Febrile**	**Defervescence**	**↑ risk**	**↓ risk**		
					Type 2 cytokine	
IL-4	•	•	•		–	Bozza, Cruz et al., 2008; Houghton-Trivino, Salgado et al., 2010; Butthep, Chunhakan et al., 2012; Kumar, Liang et al., 2012
IL-10	•	•	•		SD patients had higher IL-10 than non-severe dengue patients. IL-10 level was significantly higher in non-survival.	Bozza, Cruz et al., 2008; Houghton-Trivino, Salgado et al., 2010; Butthep, Chunhakan et al., 2012; Chunhakan, Butthep et al., 2015; Malavige, Gomes et al., 2013; Liao, Tang et al., 2015; Singla, Kar et al., 2016; Zhao, Huang et al., 2016; Abhishek, Chakravarti et al., 2017

Marker	Marker measured at		Associated with SD		Remark	Reference
	Febrile	Defervescence	↑ risk	↓ risk		
Th 17 cytokine						
IL-17	•	•	•		IL-17 started to raise from day 5 onwards.	Fernando, Wijewickrama et al., 2016
Monomokine						
IL-1β	•	•	•		–	Bozza, Cruz et al., 2008
IL-18	•	•	•		Elevation of IL-18 can already be detected in the first 4 days of illness.	Mustafa, Elbishbishi et al., 2001; Pohan, Suhendro et al., 2004; Yong, Tan et al., 2017
Chemokine						
IL-8/ CXCL8	•		•		–	Houghton-Trivino, Salgado et al., 2010; Guerrero, Arrieta et al., 2013
		•	•		–	Butthep, Chunhakan et al., 2012; Pandey, Jain et al., 2015
	•	•	•		Associated with mortality	Singla, Kar et al., 2016; Zhao, Huang et al., 2016

(*Continued*)

Table 4.1 *(Continued)*

	Marker measured at		Associated with SD		Remark	Reference
Marker	**Febrile**	**Defervescence**	**↑ risk**	**↓ risk**		
IP-10/ CXCL10	•	•	•		Elevation of IP-10 can already be detected as early as day 3 of illness.	de-Oliveira-Pinto, Gandini et al., 2012; Pang, Lindblom et al., 2016; Zhao, Huang et al., 2016
MIF	•	•	•		–	Chen, Lei et al., 2006; Yong, Tan et al., 2017
MCP-1	•	•	•		–	Singla, Kar et al., 2016; Zhao, Huang et al., 2016
MIP-1β/ CCL4	•	•		•	–	Bozza, Cruz et al., 2008; de-Oliveira-Pinto, Marinho et al., 2012
	•		•		–	Pang, Lindblom et al., 2016

Note: ↑ – increased risk; ↓ – reduced risk; IFNγ – interferon-gamma; IFNα – interferon-alpha; TNFα – tumor necrosis factor; MIF – macrophage migration inhibitory factor; MCP – Monocyte chemoattractant protein; MIP – Macrophage inflammatory proteins.

macrophages, dendritic cells, keratinocytes, endothelial cells, and platelets (Shrivastava, Valenzuela Leon et al., 2020). An increase in the mass of dengue-infected cells will therefore cause a rise in the activation of the NLRP3 inflammasome, hence more IL-1β and IL-18 being produced, culminating in the cytokine storm.

Most of the studied cytokines, including CCL2, CCL5, CCL20, CD25, CXCL1, CXCL5, CXCL9, CXCL10, CXCL11, IL-18, TNF-α, and VEGF-A, were highly expressed in dengue patients as compared to healthy individuals. The mean concentration of CCL2, CCL20, CXCL1, CXCL6, CXCL9, and TNF-α was higher in patients with SD, while CCL5, CXCL10, CXCL11, and IL-18 were expressed at lower levels in patients with SD; however, the differences were not significant. HMGB-1 (high mobility group box 1) is a DNA-binding protein that regulates innate immunity and partially adaptive immunity (Andersson and Tracey, 2011). This protein can be released into the extracellular environment passively upon apoptosis or necrosis of cells (Tang, Kang et al., 2010) or actively by activated macrophages and monocytes (Wang, Lu et al., 2010), which then functions as a pro-inflammatory cytokine in the inflammatory response positive feedback loop to stimulate endothelial cells for cytokine and chemokine production (Liu, Li et al., 2006). HMGB-1 had also been shown to induce the production of ICAM-1 in endothelial cells (Luo, Li et al., 2013), which was also expressed significantly higher in patients with SD. ICAM-1, an intercellular adhesion molecule, plays an important role in leukocytes transmigration at the sites of activated endothelial cells and the formation of immunological synapses during cellular immune responses (Dustin, 2014; Ramos, Bullard et al., 2014). Previous studies have reported the roles of HMGB-1 and ICAM-1 in dengue virus infection, where both proteins were released by dengue-infected endothelium or immune cells (Koraka, Murgue et al., 2004; Ong, Lee et al., 2012) and increased permeability of endothelial cells leading to vascular leakage (Clark, Manes et al., 2007; Huang, Liu et al., 2012). They were detected at a higher concentration in dengue patients (Cardier, Rivas et al., 2006; Allonso, Belgrano et al., 2012), while HMGB-1 was also detected in the peripheral organs of dengue fatal cases (Kontny, Kurane et al., 1988), which further justified the role of HMGB-1 and ICAM-1 in the progression of severe dengue. These are two possible predictors of severe dengue.

4.6 Metabolites Released During a Dengue Infection

Metabolomics study using QTOF-LCMS revealed that metabolites specifically regulated in DWOWS, DWWS, and SD, mostly interact with the lipid metabolism pathway. DENV has been shown to mediate lipid synthesis and metabolism in their replication cycles (Villareal, Rodgers et al., 2015) by taking advantage of the production of double-membrane vesicles during autophagy for efficient replication (Heaton and Randall, 2010). In patients with DWOWS, most of the metabolites are involved in fatty acid metabolism for energy generation and to create triglycerides, phospholipids, and other important membrane constituents (Calder, 2015), probably to maintain cellular processes to repair damages upon dengue virus infection (Hongwei Yao, 2017). On the other hand, patients with DWWS and SD expressed a large number of metabolites that are involved in phospholipid metabolism pathway, which regulates the formation and function of the membrane bilayer (Yang, Lee et al., 2018). Components from different classes of phospholipids, such as phosphatidylcholine, phosphatidylglycerol, and phosphatidylserine, as well as phosphatidic acid, which are precursors for other more complex phospholipids were expressed differentially in dengue patients, especially those with DWWS and SD. Deregulated phospholipid metabolism is likely to be due to the changes in exogenous intake of fatty acid from the patient's diet during infection or altered activities of lipid-metabolizing enzymes induced by DENV (Gimenez, Oliveros et al., 2011). Perturbations of the phospholipid metabolism might contribute to the destabilization of membrane permeability during dengue virus infection (Perera, Riley et al., 2012). Furthermore, patients with SD specifically expressed metabolites originated from the sphingolipid metabolism pathway, where sphingolipids are found mainly in the membranes of the brain and nerve cells. Altered sphingolipid metabolism might lead to the rearrangement of membrane components, which has been associated with various neurological diseases (Olsen and Faergeman, 2017) that might possibly link to the rare neurological complications, such as brachial neuropathy or encephalopathy observed in patients with SD (Verma, Sharma et al., 2011). Hence, of these, one predictive marker to consider is sphingolipid.

Metabolomics findings have also identified amino acids, such as L-isoleucine, L-phenylalanine, and 2-amino-methyl-1-butanol, which are essential in the biosynthesis of proteins were lacking in the sera of dengue patients, with the lowest prevalence in patients with SD. Changes in the plasma-free amino acid pattern, indicating infection-related alterations of amino acid metabolism, have been reported in dengue virus infection (Klassen, Furst et al., 2001). Besides, cell membrane-associated metabolites, such as C16 sphinganine and indoleacrylic acid, were also found to have a lower prevalence in dengue patients. The interaction of DENV with cellular membranes for viral replication during infection has been widely reported (Nemesio and Villalain, 2014). Since membrane-associated metabolites modulate membrane integrity by altering the biophysical properties of the cell membrane (Zhou and Blom, 2015; Wlodarska, Luo et al., 2017), the differential expression of these metabolites in dengue patients might contribute to the loss of endothelial barrier function leading to vascular leakage observed in dengue patients. Contrarily, some metabolites were detected at a higher prevalence in dengue patients, whereas healthy individuals expressed such metabolites at a lower occurrence or none at all. Etn-1-P-cer and palmitic amide, the membrane fatty acids in the family of sphingolipids, were detected higher in dengue patients, especially in patients with SD. Hence, the involvement of fatty acids needs further identification to determine, which plays a major role, and metabolite will be another predictive factor (Appendix 7: Metabolomics protocol).

Dengue NS1 protein is able to induce the loss of barrier function of the microvascular endothelium in a dose-dependent manner. However, the level of NS1 did not correlate with the extent of vascular leakage observed in the microvascular endothelium treated with serum samples from patients with dengue virus infection. This implies the presence of other host factors that might overshadow the direct effect of NS1 in inducing vascular leakage during dengue virus infection. Most of the cytokines that were highly expressed in dengue patients, including CCL2, CCL5, CCL20, and CXCL1, are involved in leukocyte infiltration where the rearrangement of junctional complex proteins such as ICAM-1, which was detected significantly higher in patients with SD, may lead to the disruption of inter-endothelial junctions. Altered lipid metabolism was detected in dengue patients, where severe manifestations were observed in patients with DWWS and SD. This might be associated with phospholipid metabolism, which may affect the membrane permeability of cells including the microvascular endothelium. Our study further highlighted the complexity of dengue as no single factor alone, either viral or host, can be responsible for the progression of severe dengue, nor can be used as a prognostic marker for disease severity. In this circumstance, the identification of these altered metabolites could facilitate dengue diagnosis or be used as a potential target for new therapeutic options.

4.7 Markers of Endothelial Activation and Coagulopathy

Vascular leakage and coagulopathy are the hallmarks of severe dengue disease (Basu and Chaturvedi, 2008; Trung and Wills, 2010; Malavige and Ogg, 2017). In normal physiological conditions, endothelium releases a wide range of vasoactive factors to regulate vascular tone, inflammation, cell adhesion, cell trafficking, coagulation, and permeability (Rajendran, Rengarajan et al., 2013). However, the dengue virus can disrupt various endothelial functions via the release of inflammatory mediators, coagulation, and apoptosis, causing vascular dysfunction and leading to vascular leakage (Basu and Chaturvedi, 2008). Besides, studies have shown that there is a prolongation in prothrombin time and partial thromboplastin time, and reduced fibrinogen, which correlates with the timing of the vascular leak and disease severity (Wills, Tran et al., 2009; Thakur, Chakravarti et al., 2016). Hence, it makes biological sense to target endothelial activation and coagulation markers as prognosis markers for severe dengue disease. These soluble markers are usually measured by ELISA or bead suspension array (Figure 4.3) and results are summarized in Table 4.2.

Table 4.2 Markers of endothelial activation and coagulopathy associated with severe dengue disease

Marker	Function	Comment	Reference
		Endothelial activation marker	
VEGF, VEGFR-1, and VEGFR-2	Important mediator for angiogenesis; previously known as vascular permeability factor and is a cytokine with permeability-enhancing ability.	VEGF levels correlated with the extent of plasma leakage were found higher in DHF patients than in DF patients. This increase in VEGF was associated with a decline of its soluble receptor, i.e., VEGFR-2.	Tseng, Lo et al., 2005; Srikiatkhachorn, 2009; Seet, Chow et al., 2009; Thakur, Chakravarti et al., 2016
Ang-1 and Ang-2	Ang-1 prevents VEGF from exerting its vascular permeability effects. Ang-2 is the antagonist of Ang-1.	Ang-1 levels significantly decreased in DHF patients. Ang-2 levels significantly increased in DHF patients.	Michels, van der Ven et al., 2012
VCAM-1 and ICAM-1	Cell adhesion molecule expressed on endothelial cells. The presence in both membrane-bound form and soluble form. The soluble forms are derived from the shedding of the membrane-bound form upon stimulation by TNFα. The level of soluble VCAM-1 and ICAM-1 indicates the level of activation of the endothelial cell.	VCAM-1 was found elevated in DSS patients compared to acute dengue fever patients. Results of ICAM-1 are thus far inconsistent, but a recent study showed that ICAM-1 was significantly elevated in severe dengue and dengue with warning signs as compared to dengue without warning signs.	Cybulsky, Iiyama et al., 2001; Koraka, Murgue et al., 2004; Khongphatthanayothin, Phumaphuti et al., 2006; Soe, Manikam et al., 2020

Table 4.2 (*Continued*)

Marker	Function	Comment	Reference
Coagulopathy marker			
Thrombomodulin	Present in large quantities on the surface of endothelial cells; acts as an anticoagulant.	Increased levels of soluble thrombomodulin are associated with DSS.	Butthep, Chunhakan et al., 2006; Sosothikul, Seksarn et al., 2007; Chen, Shyu et al., 2009
vWF	A glycoprotein synthesized by endothelial cells. Responsible for mediation of platelet adhesion at sites of endothelial injury and promoting coagulation.	Increased levels were found in DHF and DSS.	Sosothikul, Seksarn et al., 2007; Page and Liles, 2013; Djamiatun, van der Ven et al., 2012

Note: VEGF – vascular endothelial growth factor; VEGFR – vascular endothelial growth factor receptor; Ang – Angiopoietin; VCAM – vascular cell adhesion molecule; ICAM – Intercellular Adhesion Molecule 1; vWF – von Willebrand factor.

4.8 Limitations

Though measuring soluble markers in the blood are able to predict the severity of dengue disease, there are, however, several limitations. Firstly, as these soluble markers (i.e., cytokines, inflammatory mediators, vasoactive agents, and metabolites) are produced by a wide range of cells, under a wide range of biological processes, or induced by different pathogens, the disease specificity is questionable. These soluble markers alone cannot be used for the dengue diagnosis and must be used in conjunction with other microbiology, serology, or molecular biology-based diagnostic methods. Secondly, many factors can influence the levels of these soluble markers, including age, gender, host genetic factors, and days of disease onset, making interpretation of these soluble markers challenging. Careful standardization of the sample collection and patients' clinical data will be crucial for later analysis and interpretation.

Diagnosis begins with a clinical suspicion, which is usually symptoms not unique to dengue and often seen with other febrile illnesses and hence limited usefulness for the early diagnosis. Clinical symptoms have high sensitivities but have poor specificities and hence a confirmatory diagnosis is essential. The current markers used are NS1, IgM, and IgG antibodies. Virus isolation is not routinely done except in reference centers and research laboratories. Nucleic acid amplification is not as popular and only utilized in laboratories tracking infections or public health laboratories. Most of them amplify the E regions while may also include the pre-M, capsid, and NS5 regions. The titers of the IgM and IgG antibodies depend on whether the infection is a primary or secondary infection. During the primary (first) dengue infection, IgM levels are very high, but during the secondary infection, IgM levels are lower. The levels of IgG actually increase during the secondary infection. Clinicians, therefore, need to measure IgM and IgG quantitatively in order to determine whether a patient has a primary or secondary dengue infection. The main issue with confirmatory assays is the validation of each assay. Hence, the early diagnosis remains a challenge, more so the markers to use and the time frame of the detection. Using a combination of clinical markers and laboratory-confirmed assays may serve as predictive markers at different stages of the disease.

4.9 Research Aspects Needed

Serological-based assays are currently still the most popular and despite many efforts to create a single assay for confirmation, the goal has not been reached. Each diagnostic assay has wide variations in accuracy as this depends on the method applied and the epidemiological context. For appropriate usage, an understanding of the clinical conditions of dengue patients and the pattern of the immune response is essential. Assays need to be sensitive, detectable as early as possible after the onset of fever and have minimal cross-reactivity with other circulating flaviviruses. Future dengue assays have to be able to not only confirm the serotype of the virus but also have markers that are able to predict disease severity and indicate the level of immunity/protection. Biosensor assays may be the solution but must be able to be made user-friendly, cheap, and be validated in different geographical regions, as assays do not necessarily perform similarly in all countries and need to be tweaked to perform for that region where dengue is endemic. There will be high pre-test odds for the presence of anti-dengue IgG or previous infection with other flaviviruses that may interfere with the proper interpretation of test results. With these newer platforms, the usage of smaller sample volumes may assist in improving the current dengue diagnostics to one that is definitive and needs to include clinical parameters as well. One other way is the use of artificial intelligence software so that these parameters can be captured for all patients with severe dengue and would enable better monitoring and management.

References

Abhishek, K. S., A. Chakravarti, C. P. Baveja, N. Kumar, O. Siddiqui, and S. Kumar (2017). "Association of interleukin-2, -4 and -10 with dengue severity." *Indian J Pathol Microbiol*, **60**(1): 66–69.

Allonso, D., F. S. Belgrano, N. Calzada, M. G. Guzman, S. Vazquez, and R. Mohana-Borges (2012). "Elevated serum levels of high mobility group box 1 (HMGB1) protein in dengue-infected patients are associated with disease symptoms and secondary infection." *J Clin Virol*, **55**(3): 214–219.

Andersson, U. and K. J. Tracey (2011). "HMGB1 is a therapeutic target for sterile inflammation and infection." *Annu Rev Immunol*, **29**: 139–162.

Appanna, R., T. L. Huat, L. L. See, P. L. Tan, J. Vadivelu, and S. Devi (2007). "Cross-reactive T-cell responses to the nonstructural regions of dengue viruses among dengue fever and dengue hemorrhagic fever patients in Malaysia." *Clin Vaccine Immunol*, **14**(8): 969–977.

Arias, J., N. Valero, J. Mosquera, M. Montiel, E. Reyes, Y. Larreal, and M. Alvarez-Mon (2014). "Increased expression of cytokines, soluble cytokine receptors, soluble apoptosis ligand and apoptosis in dengue." *Virology*, **452–453**: 42–51.

Azeredo, E. L., L. M. De Oliveira-Pinto, S. M. Zagne, D. I. Cerqueira, R. M. Nogueira, and C. F. Kubelka (2006). "NK cells, displaying early activation, cytotoxicity and adhesion molecules, are associated with mild dengue disease." *Clin Exp Immunol*, **143**(2): 345–356.

Basu, A. and U. C. Chaturvedi (2008). "Vascular endothelium: The battlefield of dengue viruses." *FEMS Immunol Med Microbiol.* **53**(3): 287–299.

Bozza, F. A., O. G. Cruz, S. M. Zagne, E. L. Azeredo, R. M. Nogueira, E. F. Assis, P. T. Bozza, and C. F. Kubelka (2008). "Multiplex cytokine profile from dengue patients: MIP-1beta and IFN-gamma as predictive factors for severity." *BMC Infect Dis*, **8**.

Bozza, F. A., O. G. Cruz, S. M. Zagne, E. L. Azeredo, R. M. Nogueira, E. F. Assis, P. T. Bozza, and C. F. Kubelka (2008). "Multiplex cytokine profile from dengue patients: MIP-1beta and IFN-gamma as predictive factors for severity." *BMC Infect Dis*, **8**: 86.

Butthep, P., S. Chunhakan, K. Tangnararatchakit, S. Yoksan, K. Pattanapanyasat, and A. Chuansumrit (2006). "Elevated soluble thrombomodulin in the febrile stage related to patients at risk for dengue shock syndrome." *Pediatr Infect Dis J*, **25**(10): 894–897.

Butthep, P., S. Chunhakan, S. Yoksan, K. Tangnararatchakit, and A. Chuansumrit (2012). "Alteration of cytokines and chemokines during febrile episodes associated with endothelial cell damage and plasma leakage in dengue hemorrhagic fever." *Pediatr Infect Dis J*, **31**.

Calder, P. C. (2015). "Functional roles of fatty acids and their effects on human health." *JPEN J Parenter Enteral Nutr*, **39**(1): 18S–32S.

Cardier, J. E., B. Rivas, E. Romano, A. L. Rothman, C. Perez-Perez, M. Ochoa, A. M. Caceres, M. Cardier, N. Guevara, and R. Giovannetti (2006). "Evidence of vascular damage in dengue disease: demonstration of high levels of soluble cell adhesion molecules and circulating endothelial cells." *Endothelium*, **13**(5): 335–340.

Carrasco, L. R., Y. S. Leo, A. R. Cook, V. J. Lee, T. L. Thein, C. J. Go, and D. C. Lye (2014). "Predictive tools for severe dengue conforming to World Health Organization 2009 criteria." *PLoS Negl Trop Dis*, **8**(7): e2972.

Caspi, R., R. Billington, L. Ferrer, H. Foerster, C. A. Fulcher, I. M. Keseler, A. Kothari, M. Krummenacker, M. Latendresse, L. A. Mueller, Q. Ong, S. Paley, P. Subhraveti, D. S. Weaver, and P. D. Karp (2016). "The MetaCyc database of metabolic pathways and enzymes and the BioCyc collection of pathway/genome databases." *Nucleic Acids Research*, **44**(D1): D471–D480.

Chandele, A., J. Sewatanon, S. Gunisetty, M. Singla, N. Onlamoon, R. S. Akondy, H. T. Kissick, K. Nayak, E. S. Reddy, H. Kalam, D. Kumar, A. Verma, H. Panda, S. Wang, N. Angkasekwinai, K. Pattanapanyasat, K. Chokephaibulkit, G. R. Medigeshi, R. Lodha, S. Kabra, R. Ahmed, and K. Murali-Krishna (2016). "Characterization of human CD8 T cell responses in dengue virus-infected patients from India." *J Virol*, **90**(24): 11259–11278.

Chen, C. M., K. S. Chan, W. L. Yu, K. C. Cheng, H. C. Chao, C. Y. Yeh, and C. C. Lai (2016). "The outcomes of patients with severe dengue admitted to intensive care units." *Medicine (Baltimore)*, **95**(31): e4376.

Chen, L. C., H. Y. Lei, C. C. Liu, S. C. Shiesh, S. H. Chen, H. S. Liu, Y. S. Lin, S. T. Wang, H. W. Shyu, and T. M. Yeh (2006). "Correlation of serum levels of macrophage migration inhibitory factor with disease severity and clinical outcome in dengue patients." *Am J Trop Med Hyg*, **74**.

Chen, L. C., H. W. Shyu, H. M. Lin, H. Y. Lei, Y. S. Lin, H. S. Liu, and T. M. Yeh (2009). "Dengue virus induces thrombomodulin expression in human endothelial cells and monocytes in vitro." *J Infect*, **58**(5): 368–374.

Chen, R. F., J. W. Liu, W. T. Yeh, L. Wang, J. C. Chang, H. R. Yu, J. T. Cheng, and K. D. Yang (2005). "Altered T helper 1 reaction but not increase of virus load in patients with dengue hemorrhagic fever." *FEMS Immunol Med Microbiol*, **44**(1): 43–50.

Chen, R. F., K. D. Yang, L. Wang, J. W. Liu, C. C. Chiu, and J. T. Cheng (2007). "Different clinical and laboratory manifestations between dengue haemorrhagic fever and dengue fever with bleeding tendency." *Trans R Soc Trop Med Hyg*, **101**(11): 1106–1113.

Chunhakan, S., P. Butthep, S. Yoksan, K. Tangnararatchakit, and A. Chuansumrit (2015). "Vascular leakage in dengue hemorrhagic Fever is associated with dengue infected monocytes, monocyte activation/exhaustion, and cytokines production." *Int J Vasc Med*, **2015**: 917143.

Ciborowski, M., J. Teul, J. L. Martin-Ventura, J. Egido, and C. Barbas (2012). "Metabolomics with LC-QTOF-MS permits the prediction of disease stage in aortic abdominal aneurysm based on plasma metabolic fingerprint." *PLoS ONE*, **7**(2): e31982.

Clark, P. R., T. D. Manes, J. S. Pober, and M. S. Kluger (2007). "Increased ICAM-1 expression causes endothelial cell leakiness, cytoskeletal reorganization and junctional alterations." *J Invest Dermatol*, **127**(4): 762–774.

Cybulsky, M. I., K. Iiyama, H. Li, S. Zhu, M. Chen, M. Iiyama, V. Davis, J. C. Gutierrez-Ramos, P. W. Connelly, and D. S. Milstone (2001). "A major role for VCAM-1, but not ICAM-1, in early atherosclerosis." *J Clin Invest*, **107**(10): 1255–1262.

de-Oliveira-Pinto, L. M., M. Gandini, L. P. Freitas, M. M. Siqueira, C. F. Marinho, S. Setubal, C. F. Kubelka, O. G. Cruz, and S. A. Oliveira (2012). "Profile of circulating levels of IL-1Ra, CXCL10/IP-10, CCL4/MIP-1beta and CCL2/MCP-1 in dengue fever and parvovirosis." *Mem Inst Oswaldo Cruz*, **107**(1): 48–56.

de-Oliveira-Pinto, L. M., C. F. Marinho, T. F. Povoa, E. L. de Azeredo, L. A. de Souza, L. D. Barbosa, A. R. Motta-Castro, A. M. Alves, C. A. Avila, L. J. de Souza, R. V. da Cunha, P. V. Damasco, M. V. Paes, and C. F. Kubelka (2012). "Regulation of inflammatory chemokine receptors on blood T cells associated to the circulating versus liver chemokines in dengue fever." *PLoS One*, **7**(7): e38527.

de Matos, A. M., K. I. Carvalho, D. S. Rosa, L. S. Villas-Boas, W. C. da Silva, C. L. Rodrigues, O. M. Oliveira, J. E. Levi, E. S. Araujo, C. S. Pannuti, E. J. Luna, and E. G. Kallas (2015). "CD8+ T lymphocyte expansion, proliferation and activation in dengue fever." *PLoS Negl Trop Dis*, **9**(2): e0003520.

Djamiatun, K., A. J. van der Ven, P. G. de Groot, S. M. Faradz, D. Hapsari, W. M. Dolmans, S. Sebastian, R. Fijnheer, and Q. de Mast (2012). "Severe dengue is associated with consumption of von Willebrand factor and its cleaving enzyme ADAMTS-13." *PLoS Negl Trop Dis*, **6**(5): e1628.

Duong, V., S. Ly, P. Lorn Try, A. Tuiskunen, S. Ong, N. Chroeung, A. Lundkvist, I. Leparc-Goffart, V. Deubel, S. Vong, and P. Buchy (2011). "Clinical and virological factors influencing the performance of a NS1 antigen-capture assay and potential use as a marker of dengue disease severity." *PLoS Negl Trop Dis*, **5**(7): e1244.

Dustin, M. L. (2014). "The immunological synapse." *Cancer Immunol Res*, **2**(11): 1023–1033.

Duyen, H. T., T. V. Ngoc, T. Ha do, V. T. Hang, N. T. Kieu, P. R. Young, J. J. Farrar, C. P. Simmons, M. Wolbers, and B. A. Wills (2011). "Kinetics of plasma viremia and soluble nonstructural protein 1 concentrations in dengue: Differential effects according to serotype and immune status." *J Infect Dis*, **203**(9): 1292–1300.

Fernando, S., A. Wijewickrama, L. Gomes, C. T. Punchihewa, S. D. Madusanka, H. Dissanayake, C. Jeewandara, H. Peiris, G. S. Ogg, and G. N. Malavige

(2016). "Patterns and causes of liver involvement in acute dengue infection." *BMC Infect Dis*, **16**: 319.

Gandini, M., F. Petitinga-Paiva, C. F. Marinho, G. Correa, L. M. De Oliveira-Pinto, L. J. de Souza, R. V. Cunha, C. F. Kubelka, and E. L. de Azeredo (2017). "Dengue virus induces NK cell activation through TRAIL expression during infection." *Mediators Inflamm*, **2017**: 5649214.

Gimenez, M. S., L. B. Oliveros, and N. N. Gomez (2011). "Nutritional deficiencies and phospholipid metabolism." *Int J Mol Sci*, **12**(4): 2408–2433.

Green, S., S. Pichyangkul, D. W. Vaughn, S. Kalayanarooj, S. Nimmannitya, A. Nisalak, I. Kurane, A. L. Rothman, and F. A. Ennis (1999). "Early CD69 expression on peripheral blood lymphocytes from children with dengue hemorrhagic fever." *J Infect Dis*, **180**(5): 1429–1435.

Green, S., D. W. Vaughn, S. Kalayanarooj, S. Nimmannitya, S. Suntayakorn, A. Nisalak, R. Lew, B. L. Innis, I. Kurane, A. L. Rothman, and F. A. Ennis (1999). "Early immune activation in acute dengue illness is related to development of plasma leakage and disease severity." *J Infect Dis*, **179**(4): 755–762.

Guerrero, C. D., A. F. Arrieta, N. D. Ramirez, L. S. Rodriguez, R. Vega, I. Bosch, J. A. Rodriguez, C. F. Narvaez, and D. M. Salgado (2013). "High plasma levels of soluble ST2 but not its ligand IL-33 is associated with severe forms of pediatric dengue." *Cytokine*, **61**(3): 766–771.

Heaton, N. S. and G. Randall (2010). "Dengue virus-induced autophagy regulates lipid metabolism." *Cell Host Microbe*, **8**(5): 422–432.

Hongwei Yao, H., Abigail L. Peterson, and Phyllis A. Dennery (2017). "268 - role of fatty acid oxidation in hyperoxia-induced apoptosis in neonatal lung endothelial cells: Implications for lung injury and repair." *Free Radical Biology and Medicine*, **112**: 179.

Houghton-Trivino, N., D. M. Salgado, J. A. Rodriguez, I. Bosch, and J. E. Castellanos (2010). "Levels of soluble ST2 in serum associated with severity of dengue due to tumour necrosis factor alpha stimulation." *J Gen Virol*, **91**(3): 697–706.

Houng, H. S., R. Chung-Ming Chen, D. W. Vaughn, and N. Kanesa-Thasan (2001). "Development of a fluorogenic RT-PCR system for quantitative identification of dengue virus serotypes 1–4 using conserved and serotype-specific 3′ noncoding sequences." *J Virol Methods*, **95**(1–2): 19–32.

Huang, S. W., H. P. Tsai, S. J. Hung, W. C. Ko, and J. R. Wang (2020). "Assessing the risk of dengue severity using demographic information and laboratory test results with machine learning." *PLoS Negl Trop Dis*, **14**(12): e0008960.

Huang, W., Y. Liu, L. Li, R. Zhang, W. Liu, J. Wu, E. Mao, and Y. Tang (2012). "HMGB1 increases permeability of the endothelial cell monolayer via RAGE and Src family tyrosine kinase pathways." *Inflammation*, **35**(1): 350–362.

Jaiyen, Y., P. Masrinoul, S. Kalayanarooj, R. Pulmanausahakul, and S. Ubol (2009). "Characteristics of dengue virus-infected peripheral blood mononuclear cell death that correlates with the severity of illness." *Microbiol Immunol*, **53**(8): 442–450.

Jenkins, S., S. Fischer, and T. Sana (2013). "Compound identification, profiling and pathway analysis of the yeast metabolome in mass profiler professional." Santa Clara, CA, USA, Agilent Technologies, Inc.

Keawvichit, R., L. Khowawisetsut, S. Lertjuthaporn, K. Tangnararatchakit, N. Apiwattanakul, S. Yoksan, A. Chuansumrit, K. Chokephaibulkit, A. A. Ansari, N. Onlamoon, and K. Pattanapanyasat (2018). "Differences in activation and tissue homing markers of natural killer cell subsets during acute dengue infection." *Immunology*, **153**(4): 455–465.

Khongphatthanayothin, A., P. Phumaphuti, K. Thongchaiprasit, and Y. Poovorawan (2006). "Serum levels of sICAM-1 and sE-selectin in patients with dengue virus infection." *Jpn J Infect Dis.* **59**(3): 186–188.

Kittitrakul, C., U. Silachamroon, W. Phumratanaprapin, S. Krudsood, P. Wilairatana, and S. Treeprasertsuk (2015). "Liver function tests abnormality and clinical severity of dengue infection in adult patients." *J Med Assoc Thai*, **98**(1): S1–8.

Klassen, P., P. Furst, C. Schulz, M. Mazariegos, and N. W. Solomons (2001). "Plasma free amino acid concentrations in healthy Guatemalan adults and in patients with classic dengue." *Am J Clin Nutr*, **73**(3): 647–652.

Kong, Y. Y., C. H. Thay, T. C. Tin, and S. Devi (2006). "Rapid detection, serotyping and quantitation of dengue viruses by TaqMan real-time one-step RT-PCR." *J Virol Methods*, **138**(1–2): 123–130.

Kontny, U., I. Kurane, and F. A. Ennis (1988). "Gamma interferon augments Fc gamma receptor-mediated dengue virus infection of human monocytic cells." *J Virol*, **62**(11): 3928–3933.

Koraka, P., B. Murgue, X. Deparis, E. C. Van Gorp, T. E. Setiati, A. D. Osterhaus, and J. Groen (2004). "Elevation of soluble VCAM-1 plasma levels in children with acute dengue virus infection of varying severity." *J Med Virol*, **72**(3): 445–450.

Kuczera, D., J. P. Assolini, F. Tomiotto-Pellissier, W. R. Pavanelli, and G. F. Silveira (2018). "Highlights for dengue immunopathogenesis: Antibody-dependent enhancement, cytokine storm, and beyond." *J Interferon Cytokine Res*, **38**(2): 69–80.

Kumar, Y., C. Liang, Z. Bo, J. C. Rajapakse, E. E. Ooi, and S. R. Tannenbaum (2012). "Serum proteome and cytokine analysis in a longitudinal cohort of adults with primary dengue infection reveals predictive markers of DHF." *PLoS Negl Trop Dis*, **6**.

Lee, L. K., V. C. Gan, V. J. Lee, A. S. Tan, Y. S. Leo, and D. C. Lye (2012). "Clinical relevance and discriminatory value of elevated liver aminotransferase levels for dengue severity." *PLoS Negl Trop Dis*, **6**(6): e1676.

Lei, H. Y., T. M. Yeh, H. S. Liu, Y. S. Lin, S. H. Chen, and C. C. Liu (2001). "Immunopathogenesis of dengue virus infection." *J Biomed Sci*, **8**(5): 377–388.

Leo, Y. S., V. C. Gan, E. L. Ng, Y. Hao, L. C. Ng, K. Y. Pok, F. Dimatatac, C. J. Go, and D. C. Lye (2013). "Utility of warning signs in guiding admission and predicting severe disease in adult dengue." *BMC Infect Dis*, **13**: 498.

Liao, B., Y. Tang, F. Hu, W. Zhou, X. Yao, W. Hong, J. Wang, X. Zhang, X. Tang, and F. Zhang (2015). "Serum levels of soluble vascular cell adhesion molecules may correlate with the severity of dengue virus-1 infection in adults." *Emerg Microbes Infect*, **4**(4): e24.

Libraty, D. H., T. P. Endy, H. S. Houng, S. Green, S. Kalayanarooj, S. Suntayakorn, W. Chansiriwongs, D. W. Vaughn, A. Nisalak, F. A. Ennis, and A. L. Rothman (2002). "Differing influences of virus burden and immune activation on disease severity in secondary dengue-3 virus infections." *J Infect Dis*, **185**(9): 1213–1221.

Libraty, D. H., P. R. Young, D. Pickering, T. P. Endy, S. Kalayanarooj, S. Green, D. W. Vaughn, A. Nisalak, F. A. Ennis, and A. L. Rothman (2002). "High circulating levels of the dengue virus nonstructural protein NS1 early in dengue illness correlate with the development of dengue hemorrhagic fever." *J Infect Dis*, **186**(8): 1165–1168.

Littaua, R., I. Kurane, and F. A. Ennis (1990). "Human IgG Fc receptor II mediates antibody-dependent enhancement of dengue virus infection." *J Immunol*, **144**(8): 3183–3186.

Liu, J. H., Z. J. Li, J. Tang, Y. W. Liu, L. Zhao, P. Deng, and Y. Jiang (2006). "High mobility group box-1 protein activates endothelial cells to produce cytokines and has synergistic effect with lipopolysaccharide in inducing interleukin-6 release." *Zhonghua Yi Xue Za Zhi*, **86**(17): 1191–1195.

Luetke-Eversloh, M., M. Killig, and C. Romagnani (2013). "Signatures of human NK cell development and terminal differentiation." *Front Immunol*, **4**: 499.

Luo, Y., S. J. Li, J. Yang, Y. Z. Qiu, and F. P. Chen (2013). "HMGB1 induces an inflammatory response in endothelial cells via the RAGE-dependent endoplasmic reticulum stress pathway." *Biochem Biophys Res Commun*, **438**(4): 732–738.

Malavige, G. N., L. Gomes, L. Alles, T. Chang, M. Salimi, S. Fernando, K. D. Nanayakkara, S. Jayaratne, and G. S. Ogg (2013). "Serum IL-10 as a marker of severe dengue infection." *BMC Infectious Diseases*, **13**(1): 341.

Malavige, G. N. and G. S. Ogg (2017). "Pathogenesis of vascular leak in dengue virus infection." *Immunology*, **151**(3): 261–269.

Messina, J. P., O. J. Brady, T. W. Scott, C. Zou, D. M. Pigott, K. A. Duda, S. Bhatt, L. Katzelnick, R. E. Howes, K. E. Battle, C. P. Simmons, and S. I. Hay (2014). "Global spread of dengue virus types: Mapping the 70 year history." *Trends Microbiol*, **22**(3): 138–146.

Michels, M., Q. de Mast, M. G. Netea, L. A. Joosten, C. A. Dinarello, P. I. Rudiman, S. Sinarta, R. Wisaksana, B. Alisjahbana, and A. J. van der Ven (2015). "Normal free interleukin-18 (IL-18) plasma levels in dengue virus infection and the need to measure both total IL-18 and IL-18 binding protein levels." *Clin Vaccine Immunol*, **22**(6): 650–655.

Michels, M., A. J. van der Ven, K. Djamiatun, R. Fijnheer, P. G. de Groot, A. W. Griffioen, S. Sebastian, S. M. Faradz, and Q. de Mast (2012). "Imbalance of angiopoietin-1 and angiopoetin-2 in severe dengue and relationship with thrombocytopenia, endothelial activation, and vascular stability." *Am J Trop Med Hyg*, **87**(5): 943–946.

Moi, M. L., C. K. Lim, T. Takasaki, and I. Kurane (2010). "Involvement of the Fc gamma receptor IIA cytoplasmic domain in antibody-dependent enhancement of dengue virus infection." *J Gen Virol*, **91**(1): 103–111.

Mustafa, A. S., E. A. Elbishbishi, R. Agarwal, and U. C. Chaturvedi (2001). "Elevated levels of interleukin-13 and IL-18 in patients with dengue hemorrhagic fever." *FEMS Immunol Med Microbiol*, **30**(3): 229–233.

Nanda, J. D., T. S. Ho, R. D. Satria, M. K. Jhan, Y. T. Wang, and C. F. Lin (2021). "IL-18: The forgotten cytokine in dengue immunopathogenesis." *J Immunol Res*, **2021**: 8214656.

Nemesio, H. and J. Villalain (2014). "Membrane interacting regions of dengue virus NS2A protein." *J Phys Chem B*, **118**(34): 10142–10155.

Olsen, A. S. B. and N. J. Faergeman (2017). "Sphingolipids: membrane microdomains in brain development, function and neurological diseases." *Open Biol*, **7**(5).

Ong, S. P., L. M. Lee, Y. F. Leong, M. L. Ng, and J. J. Chu (2012). "Dengue virus infection mediates HMGB1 release from monocytes involving PCAF acetylase complex and induces vascular leakage in endothelial cells." *PLoS One*, **7**(7): e41932.

Padyana, M., S. Karanth, S. Vaidya, and J. A. Gopaldas (2019). "Clinical profile and outcome of dengue fever in multidisciplinary intensive care unit of a tertiary level hospital in India." *Indian J Crit Care Med*, **23**(6): 270–273.

Page, A. V. and W. C. Liles (2013). "Biomarkers of endothelial activation/dysfunction in infectious diseases." *Virulence*, **4**(6): 507–516.

Pan, P., Q. Zhang, W. Liu, W. Wang, Z. Lao, W. Zhang, M. Shen, P. Wan, F. Xiao, F. Liu, W. Zhang, Q. Tan, X. Liu, K. Wu, Y. Liu, G. Li, and J. Wu (2019). "Dengue virus M protein promotes NLRP3 inflammasome activation to induce vascular leakage in mice." *J Virol*, **93**(21).

Pandey, N., A. Jain, R. K. Garg, R. Kumar, O. P. Agrawal, and P. V. Lakshmana Rao (2015). "Serum levels of IL-8, IFNgamma, IL-10, and TGF beta and their gene expression levels in severe and non-severe cases of dengue virus infection." *Arch Virol*, **160**(6): 1463–1475.

Pang, J., A. Lindblom, T. Tolfvenstam, T. L. Thein, A. N. Naim, L. Ling, A. Chow, M. I. Chen, E. E. Ooi, Y. S. Leo, and M. L. Hibberd (2016). "Discovery and validation of prognostic biomarker models to guide triage among adult dengue patients at early infection." *PLoS One*, **11**(6): e0155993.

Paranavitane, S. A., L. Gomes, A. Kamaladasa, T. N. Adikari, N. Wickramasinghe, C. Jeewandara, N. L. Shyamali, G. S. Ogg, and G. N. Malavige (2014). "Dengue NS1 antigen as a marker of severe clinical disease." *BMC Infect Dis*, **14**: 570.

Parsa, S. F., A. Vafajoo, A. Rostami, R. Salarian, M. Rabiee, N. Rabiee, G. Rabiee, M. Tahriri, A. Yadegari, D. Vashaee, L. Tayebi, and M. R. Hamblin (2018). "Early diagnosis of disease using microbead array technology: A review." *Anal Chim Acta*, **1032**: 1–17.

Pawitan, J. A. (2011). "Dengue virus infection: predictors for severe dengue." *Acta Med Indones*, **43**(2): 129–135.

Perera, R., C. Riley, G. Isaac, A. S. Hopf-Jannasch, R. J. Moore, K. W. Weitz, L. Pasa-Tolic, T. O. Metz, J. Adamec, and R. J. Kuhn (2012). "Dengue virus infection perturbs lipid homeostasis in infected mosquito cells." *PLoS Pathog*, **8**(3): e1002584.

Pitt, J. J. (2009). "Principles and applications of liquid chromatography-mass spectrometry in clinical biochemistry." *The Clinical Biochemist Reviews*, **30**(1): 19–34.

Pohan, H. T., S. Suhendro, R. Bur, A. Matondang, S. Djauzi, K. Inada, and S. Endo (2004). "Interleukin-18 levels in adult dengue fever and dengue hemorrhagic fever." **13**(2): 76.

Puc, I., T. C. Ho, K. L. Yen, A. Vats, J. J. Tsai, P. L. Chen, Y. W. Chien, Y. C. Lo, and G. C. Perng (2021). "Cytokine signature of dengue patients at different severity of the disease." *Int J Mol Sci*, **22**(6).

Rajendran, P., T. Rengarajan, J. Thangavel, Y. Nishigaki, D. Sakthisekaran, G. Sethi, and I. Nishigaki (2013). "The vascular endothelium and human diseases." *Int J Biol Sci*, **9**(10): 1057–1069.

Ramos, T. N., D. C. Bullard, and S. R. Barnum (2014). "ICAM-1: Isoforms and phenotypes." *J Immunol*, **192**(10): 4469–4474.

Rathakrishnan, A., B. Klekamp, S. M. Wang, T. V. Komarasamy, S. K. Natkunam, J. Sathar, A. Azizan, A. Sanchez-Anguiano, R. Manikam, and S. D. Sekaran (2014). "Clinical and immunological markers of dengue progression in a study cohort from a hyperendemic area in Malaysia." *PLoS One*, **9**(3): e92021.

Rodrigo, W. W., X. Jin, S. D. Blackley, R. C. Rose, and J. J. Schlesinger (2006). "Differential enhancement of dengue virus immune complex infectivity mediated by signaling-competent and signaling-incompetent human Fcgamma RIA (CD64) or FcgammaRIIA (CD32)." *J Virol*, **80**(20): 10128–10138.

Rothman, A. L. (2011). "Immunity to dengue virus: a tale of original antigenic sin and tropical cytokine storms." *Nat Rev Immunol*, **11**(8): 532–543.

Seet, R. C., A. W. Chow, A. M. Quek, Y. H. Chan, and E. C. Lim (2009). "Relationship between circulating vascular endothelial growth factor and its soluble receptors in adults with dengue virus infection: A case-control study." *Int J Infect Dis*, **13**(5): e248–253.

Senaratne, T., J. Carr, and F. Noordeen (2016). "Elevation in liver enzymes is associated with increased IL-2 and predicts severe outcomes in clinically apparent dengue virus infection." *Cytokine*, **83**: 182–188.

Shrivastava, G., P. C. Valenzuela Leon, and E. Calvo (2020). "Inflammasome fuels dengue severity." *Front Cell Infect Microbiol*, **10**: 489.

Silveira, G. F., P. F. Wowk, A. H. D. Cataneo, P. F. Dos Santos, M. Delgobo, M. A. Stimamiglio, M. Lo Sarzi, A. Thomazelli, I. Conchon-Costa, W. R. Pavanelli, L. R. V. Antonelli, A. Bafica, D. S. Mansur, C. N. D. Dos Santos, and J. Bordignon (2018). "Human T lymphocytes are permissive for dengue virus replication." *J Virol*, **92**(10).

Simmons, C. P., J. J. Farrar, V. Nguyen, and B. Wills (2012). "Dengue." *N Engl J Med*, **366**(15): 1423–1432.

Singla, M., M. Kar, T. Sethi, S. K. Kabra, R. Lodha, A. Chandele, and G. R. Medigeshi (2016). "Immune response to dengue virus infection in pediatric patients in New Delhi, India - Association of Viremia, Inflammatory Mediators and Monocytes with Disease Severity." *PLoS Negl Trop Dis*, **10**(3): e0004497.

Soe, H. J., R. Manikam, C. S. Raju, M. A. Khan, and S. D. Sekaran (2020). "Correlation of host inflammatory cytokines and immune-related metabolites, but not viral NS1 protein, with disease severity of dengue virus infection." *PLoS One*, **15**(8): e0237141.

Sosothikul, D., P. Seksarn, S. Pongsewalak, U. Thisyakorn, and J. Lusher (2007). "Activation of endothelial cells, coagulation and fibrinolysis in children with dengue virus infection." *Thromb Haemost*, **97**(4): 627–634.

Soundravally, R., S. L. Hoti, S. A. Patil, C. C. Cleetus, B. Zachariah, T. Kadhiravan, P. Narayanan, and B. A. Kumar (2014). "Association between proinflammatory cytokines and lipid peroxidation in patients with severe dengue disease around defervescence." *Int J Infect Dis*, **18**: 68–72.

Srikiatkhachorn, A. (2009). "Plasma leakage in dengue haemorrhagic fever." *Thromb Haemost*, **102**(6): 1042–1049.

Srikiatkhachorn, A., A. Mathew, and A. L. Rothman (2017). "Immune-mediated cytokine storm and its role in severe dengue." *Semin Immunopathol*, **39**(5): 563–574.

Tang, D., R. Kang, C. W. Cheh, K. M. Livesey, X. Liang, N. E. Schapiro, R. Benschop, L. J. Sparvero, A. A. Amoscato, K. J. Tracey, H. J. Zeh, and M. T. Lotze (2010). "HMGB1 release and redox regulates autophagy and apoptosis in cancer cells." *Oncogene*, **29**(38): 5299–5310.

Thakur, P., A. Chakravarti, S. Aggarwal, B. Uppal, and P. Bhalla (2016). "Elevated levels of vascular endothelial growth factor in adults with severe dengue infection." *Virusdisease*, **27**(1): 48–54.

Trung, D. T. and B. Wills (2010). "Systemic vascular leakage associated with dengue infections: The clinical perspective." *Curr Top Microbiol Immunol*, **338**: 57–66.

Tseng, C. S., H. W. Lo, H. C. Teng, W. C. Lo, and C. G. Ker (2005). "Elevated levels of plasma VEGF in patients with dengue hemorrhagic fever." *FEMS Immunol Med Microbiol*, **43**(1): 99–102.

van de Weg, C. A., C. S. Pannuti, E. S. de Araujo, H. J. van den Ham, A. C. Andeweg, L. S. Boas, A. C. Felix, K. I. Carvalho, A. M. de Matos, J. E. Levi, C. M. Romano, C. C. Centrone, C. L. de Lima Rodrigues, E. Luna, E. C. van Gorp, A. D. Osterhaus, B. E. Martina, and E. G. Kallas (2013). "Microbial translocation is associated with extensive immune activation in dengue virus infected patients with severe disease." *PLoS Negl Trop Dis*, **7**(5): e2236.

Vaughn, D. W., S. Green, S. Kalayanarooj, B. L. Innis, S. Nimmannitya, S. Suntayakorn, T. P. Endy, B. Raengsakulrach, A. L. Rothman, F. A. Ennis, and A. Nisalak (2000). "Dengue viremia titer, antibody response pattern, and virus serotype correlate with disease severity." *J Infect Dis*, **181**(1): 2–9.

Verma, R., P. Sharma, R. K. Garg, V. Atam, M. K. Singh, and H. S. Mehrotra (2011). "Neurological complications of dengue fever: Experience from a tertiary center of north India." *Ann Indian Acad Neurol*, **14**(4): 272–278.

Villareal, V. A., M. A. Rodgers, D. A. Costello, and P. L. Yang (2015). "Targeting host lipid synthesis and metabolism to inhibit dengue and hepatitis C viruses." *Antiviral Res*, **124**: 110–121.

Waggoner, J. J., J. Abeynayake, M. K. Sahoo, L. Gresh, Y. Tellez, K. Gonzalez, G. Ballesteros, A. M. Pierro, P. Gaibani, F. P. Guo, V. Sambri, A. Balmaseda, K. Karunaratne, E. Harris, and B. A. Pinsky (2013). "Single-reaction, multiplex, real-time rt-PCR for the detection, quantitation, and serotyping of dengue viruses." *PLoS Negl Trop Dis*, **7**(4): e2116.

Wahid, S. F., S. Sanusi, M. M. Zawawi, and R. A. Ali (2000). "A comparison of the pattern of liver involvement in dengue hemorrhagic fever with classic dengue fever." *Southeast Asian J Trop Med Public Health*, **31**(2): 259–263.

Wang, F., Z. Lu, M. Hawkes, H. Yang, K. C. Kain, and W. C. Liles (2010). "Fas (CD95) induces rapid, TLR4/IRAK4-dependent release of pro-inflammatory HMGB1 from macrophages." *J Inflamm (Lond)*, **7**: 30.

Wang, L., R. F. Chen, J. W. Liu, H. R. Yu, H. C. Kuo, and K. D. Yang (2007). "Implications of dynamic changes among tumor necrosis factor-alpha (TNF-alpha), membrane TNF receptor, and soluble TNF receptor levels in regard to the severity of dengue infection." *Am J Trop Med Hyg*, **77**(2): 297–302.

WHO (2009). *WHO (2009) Dengue Guidelines for Diagnosis, Treatment, Prevention and Control: New Edition*. WHO, Geneva.

Wills, B., V. N. Tran, T. H. Nguyen, T. T. Truong, T. N. Tran, M. D. Nguyen, V. D. Tran, V. V. Nguyen, T. T. Dinh, and J. Farrar (2009). "Hemostatic changes in Vietnamese children with mild dengue correlate with the severity of vascular leakage rather than bleeding." *Am J Trop Med Hyg*, **81**(4): 638–644.

Wlodarska, M., C. Luo, R. Kolde, E. d'Hennezel, J. W. Annand, C. E. Heim, P. Krastel, E. K. Schmitt, A. S. Omar, E. A. Creasey, A. L. Garner, S. Mohammadi, D. J. O'Connell, S. Abubucker, T. D. Arthur, E. A. Franzosa, C. Huttenhower, L. O. Murphy, H. J. Haiser, H. Vlamakis, J. A. Porter, and R. J. Xavier (2017). "Indoleacrylic acid produced by commensal peptostreptococcus species suppresses inflammation." *Cell Host Microbe*, **22**(1): 25–37 e26.

Yacoub, S., P. K. Lam, H. M. Vu le, T. L. Le, N. T. Ha, T. T. Toan, N. T. Van, N. T. Quyen, H. T. Le Duyen, N. Van Kinh, A. Fox, J. Mongkolspaya, M. Wolbers, C. P. Simmons, G. R. Screaton, H. Wertheim, and B. Wills (2016). "Association of microvascular function and endothelial biomarkers with clinical outcome in dengue: An observational study." *J Infect Dis*, **214**(5): 697–706.

Yacoub, S., J. Mongkolsapaya, and G. Screaton (2013). "The pathogenesis of dengue." *Curr Opin Infect Dis*, **26**(3): 284–289.

Yang, Y., M. Lee, and G. D. Fairn (2018). "Phospholipid subcellular localization and dynamics." *J Biol Chem*, **293**(17): 6230–6240.

Yong, Y. K., H. Y. Tan, S. H. Jen, E. M. Shankar, S. K. Natkunam, J. Sathar, R. Manikam, and S. D. Sekaran (2017). "Aberrant monocyte responses predict and characterize dengue virus infection in individuals with severe disease." *J Transl Med*, **15**(1): 121.

Yong, Y. K., R. Thayan, H. T. Chong, C. T. Tan, and S. D. Sekaran (2007). "Rapid detection and serotyping of dengue virus by multiplex RT-PCR and real-time SYBR green RT-PCR." *Singapore Med J*, **48**(7): 662–668.

Zhang, H., Y. P. Zhou, H. J. Peng, X. H. Zhang, F. Y. Zhou, Z. H. Liu, and X. G. Chen (2014). "Predictive symptoms and signs of severe dengue disease for patients with dengue fever: A meta-analysis." *Biomed Res Int*, **2014**: 359308.

Zhao, L., X. Huang, W. Hong, S. Qiu, J. Wang, L. Yu, Y. Zeng, X. Tan, and F. Zhang (2016). "Slow resolution of inflammation in severe adult dengue patients." *BMC Infect Dis*, **16**: 291.

Zhou, B., J. F. Xiao, L. Tuli, and H. W. Ressom (2012). "LC-MS-based metabolomics." *Molecular bioSystems*, **8**(2): 470–481.

Zhou, K. and T. Blom (2015). "Trafficking and functions of bioactive sphingolipids: Lessons from cells and model membranes." *Lipid Insights*, **8**(1l): 11–20.

Zompi, S. and E. Harris (2013). "Original antigenic sin in dengue revisited." *Proc Natl Acad Sci USA*, **110**(22): 8761–8762.

Chapter 5

Research and Recommendations

Shamala Devi Sekaran
Faculty of Applied Sciences, UCSI University Kuala Lumpur Campus, Malaysia
shamala@ucsiuniversity.edu.my

Diagnosis begins with a clinical suspicion, which is usually a symptom not unique to dengue and often seen with other febrile illnesses and hence limited usefulness for early diagnosis. Clinical symptoms have high sensitivities but have poor specificities and hence a confirmatory diagnosis is essential. The current markers used are NS1, IgM, and IgG antibodies. Virus isolation is not routinely done except in reference centers and research laboratories. Nucleic acid amplification is not as popular and only utilized in laboratories tracking infections or public health laboratories. Most of them amplify the E regions while may also include the pre-M, capsid, or NS5 regions. The titers of the IgM and IgG antibodies depend on whether the infection is a primary or secondary infection. During primary (first) dengue infection, IgM levels are very high, but during secondary infection, IgM levels are lower. The levels of IgG actually increase during secondary infection. Clinicians, therefore, need to

Dengue Diagnostics: The Right Test at the Right Time for the Right Group
Edited by Shamala Devi Sekaran

ISBN 978-981-4968-97-3 (Hardcover), 978-1-032-66956-4 (eBook)
www.jennystanford.com

measure IgM and IgG quantitatively in order to determine whether a patient has a primary or secondary dengue infection. The main issues with confirmatory assays are the validation of each assay. Hence, early diagnosis remains a challenge, more so the markers to use and the time frame of detection. Using a combination of clinical markers and laboratory-confirmed assays may serve as predictive markers at different stages of the disease.

5.1 Research Aspects Needed

The question to ask is what marker or markers should we be looking out for? Over the years from isolating the virus to its detection, followed by the detection of IgM, then IgG, and then total antibodies were used. Upon the advent of newer technologies, more precise assays were developed, with newer markers. However, the question still remains as many of these markers are present for too short a time or are not at the specificity needed to differentiate them from other flaviviruses. To understand this, we need to go back to how initially viral infections were diagnosed for convenience and simplicity with paired samples to note a rising immune response. Not to forget of course is that the gold standard is the presence of the virus, which although gives the greatest confidence is not convenient for many laboratories or physicians at out-patient clinics. Which marker then is the best or is there only one marker to use? We now know that the immune response is a dynamic one and the effects it has during a viral infection determine how we go about a diagnosis. Should we use NSI alone? We have found that we cannot do this as patient-to-patient variation occurs as with all other markers as well. Neither can we rely on single IgM and IgG qualitatively. At the same time, multiple clinical markers have emerged and hence the need to also consider these. Some of these include ALT/AST, platelet counts, HCT, vWF, thrombomodulin, urinary levels of heparin, plasma levels of pentraxin 3, levels of VEGF, and soluble VCAM-1. Apart from that research has shown a number of metabolites, which could be considered on a patient-to-patient basis such as those involved in lipid and phospholipid metabolism and sphingolipids in cases of encephalopathy.

Serological-based assays are currently still the most popular and despite the many efforts to create a single assay for confirmation, the goal has not been reached. Each diagnostic assay has wide variations in accuracy as this depends on the method applied and the epidemiological context. For appropriate usage, an understanding of the clinical conditions of dengue patients and the pattern of the immune response is essential. Assays need to be sensitive, detectable as early as possible after the onset of fever, and have minimal cross-reactivity with other circulating flaviviruses. Future dengue assays have to be able to not only confirm the serotype of the virus but also have markers that are able to predict disease severity and indicate a level of immunity/protection. Biosensor assays may be the solution but must be able to be made user-friendly, cheap, and be validated in different geographical regions, as assays do not necessarily perform similarly in all countries and need to be tweaked to perform for that region where dengue is endemic. There will be high pre-test odds for the presence of anti-dengue IgG or previous infection with other flaviviruses that may interfere with the proper interpretation of test results. With these newer platforms, the usage of smaller sample volumes may assist in improving the current dengue diagnostics to one that is definitive and needs to include clinical parameters as well. One other way is the use of artificial intelligence software so that these parameters can be captured over time for all patients with severe dengue and would enable better monitoring and management.

Selected Protocols

Appendix 1: Inoculation of Mosquitoes

Mosquitoes are inoculated intrathoracically with serum or plasma specimens. The quantitation of the virus in specimens is calculated using the method of Reed and Muench and expressed as the dose that infects 50% of the mosquitoes inoculated (MID_{50}). Briefly, aspirate 3–5-day-old mosquitoes and anesthetize them by holding them in tubes on ice for at least 30 min. Place 3–5 mosquitoes on a petri dish held on a tall stand under a stereoscopic microscope and position the mosquito on the dorsal aspect of its thorax. Impale the mosquito on the underside of the neck. Load the needle with inoculum and release the plunger of the syringe until the desired amount has been inoculated. Inoculate 10–20 mosquitoes to form a virus seed pool. Place the mosquito within a carton and incubate it in an insect growth chamber for 10 days at 28 °C with 80% humidity containing 10% sucrose solution.

Appendix 2: Inoculation of Cell Lines

Either mosquito (C6/36, AP-16, and Tra-284) or mammalian (Vero, BS-C-1, LLMK2, PS, and BHK 21) cell lines can be used to culture DENV from specimens. Briefly grow cells until 95% confluency in a 24 wells plate with an appropriate growth media and conditions designated for the specific cell lines. Remove the growth media and replace it with media supplemented with 1% FBS. Inoculate 5 μl of specimens (e.g., serum, cerebral spinal fluid) to the cells and incubate overnight. Check the cells the following day. Add fresh cells to each well if cells start to die due to the effects of the cytokines. Observe cells daily for 14 days under the microscope. Harvest the virus if cell death is observed.

Appendix 3: Inoculation of Suckling Mice

Inoculate BALB/c mice by the intracerebral (i.c.) route with 2,04 log_{10} PFU in 30 µl (40 LD_{50}) or 6 log_{10} virus RNA copies/ml of virus in culture medium. Sacrifice the mice on days 1, 3, 5, and 7, and endpoints post-infection (dpi). The clinical signs of infection with the commitment of the CNS generally appear after the 7th dpi, when the vertebral column is impaired (curved) and the loss of strength or paralysis is observed in the lower and upper limbs. The brain tissue exhibits thickening of the pia mater with the presence of inflammation and vascular lesions, while microglia hyperplasia in cerebral and cerebellar tissues is prominent in the white matter (Amorim et al., 2019).

Appendix 4: PCR

One-step real-time Sybr Green RT-PCR was performed using Biorad iTaq universal one-step Sybr Green premix (Biorad, Hercules, CA, USA) and in-house designed primers (Y. K. Yong, Thayan, Chong, Tan, & Sekaran, 2007) (Table A4.1).

Table A4.1 Primer sequences and the size of RT-PCR product of the combined primers

DENV serotype	Primer	Primer sequence	Size of amplicon
	DF	5′-AGT TGT TAG TCT ACG TGG ACC GAC A	
DENV1	D1	5′-CCC CGT AAC ACT TTG ATC GCT CCA TT	342 bp (DF and D1)
DENV2	D2	5′-CGC CAC AAG GGC CAT GAA CAG	251 bp (DF and D2)
DENV3	D3	5′-GCA CAT GTT GAT TCC AGA GGC TGT C	538 bp (DF and D3)
DENV4	D4	5′-GTT TCC AAT CCC ATT CCT GAA TGT GGT GT	754 bp (DF and D4)

The reagent mix for RT-PCR consists of 5 µl iTaq universal Sybr Green reaction mix, 0.15 µl iScript reverse transcriptase, 10 µm of forward and reverse primers specific to the four serotypes separately, and up to 500 ng of RNA. The thermal cycling profile consisted of a

10-min reverse transcriptase step at 50 °C, 1 min of Taq polymerase activation and DNA denaturation, and 95 °C, followed by 40 cycles of PCR at 95 °C for 10 seconds and 60 °C of annealing and extension for 30 seconds. Melting curve analysis was then performed consisting of a denaturation step at 95 °C for 1 min, lowered to 55 °C for 30 seconds, followed by 80 cycles of incubation at a rate of 0.5 °C per 10 seconds per cycle to a final temperature of 95 °C. The amplification with a threshold cycle (Ct value) lower than 35 cycles and a melting point ranging from 81 to 83 °C indicates a positive result corresponding to the specific serotype observed. No amplification or amplification with a high Ct value (> 35 cycles) and a melting point ranging from 75 to 79 °C (presence of primer dimers) indicates a negative result.

Appendix 5: Nucleic Acid Sequence–Based Amplification (NASBA)

Dengue viral RNA is amplified by NASBA according to the procedure of Usawattanakul et al., 2002. Briefly, primers and probes used are listed as follows: P1[5′-aat tct aat acg act cac tat agg gga gac (T7 promoter) AGC AGG ATC TCT GGT CT-3′], P2[5′-gat gca agg tcg cat atg agg gtt aga gga (RCL tail) GAC CCC TCC C-3′], probe [5′-AAA CAG CAT ATT GAC GCT GGG-3′]. Nucleic acid amplification is performed in a 20 μl reaction containing 40 mM Tris-HCl (pH 8.5), 12 mM $MgCl_2$, 70 mM KCl, 5 mM dithiothreitol, 1 mM of each dATP, dCTP, dGTP, and dTTP, 2 mM of each ATP, CTP, and UTP, 1.5 mM GTP, 0.5 mM ITP, 0.1 μg/μl of BSA, 1.5 M sorbitol, 15% dimethyl sulfoxide (DMSO), 0.08 U RNase-H, 32 U T7 RNA polymerase, 6.4 U avian myeloblastosis virus reverse transcriptase (AMV-RT), 0.2 μM of each primer, and 5 μl of extracted nucleic acid. For nucleic acid detection, the NASBA amplified products are diluted 1:20 in detection diluent (1.0 mM Tris-HCl (pH 8.5), 0.2 g/L methylisothiazolinone), and incubated with biotinylated dengue virus-specific probe bound to 5 μg of streptavidin-coated paramagnetic beads and 3×10^{11} molecules of the ruthenium-labeled oligonucleotides detection probe. Then 300 μl of assay buffer (100 mM tripropylamine, pH 7.5) is added before reading by an ECL reader. NASBA amplification products can also be visualized by agarose gel electrophoresis.

Appendix 6: Example of a Protocol that Utilizes Biosensors

1. *Perform nucleic acid extraction from 200 µl of a sample using the taco pre-loaded DNA/RNA extraction kit (GeneReach) on a taco mini-automatic nucleic acid extraction system (taco mini; GeneReach).*
2. *Rehydrate lyophilized RT-iiPCR reagents (POCKIT Dengue Virus Serotype 1, 2, 3, or 4 Reagent Set; GeneReach) with 50 µl of premix buffer.*
3. *Add 5 µl of sample to the mixture.*
4. *Transfer 50 µl of the final mixture to an R-tube (GeneReach), and spin briefly in a cube mini centrifuge.*
5. *Place the R-tubes into the reaction chamber of a POCKIT nucleic acid analyzer or a hand-held POCKIT micro plus nucleic acid analyzer, and initiate the run.*
6. *Apply the default program, including an RT step at 50 °C for 10 min and an iiPCR step at 95 °C for about 30 min.*

Appendix 7: CRISPR/Cpf1-Based Electrochemical Biosensor System for DENV-4 RNA Detection

1. *Prepare methylene blue (MB)-conjugated streptavidin gold nanoparticles (STV-AuNP) in phosphate-buffered saline (pH 7.4) at a molar ratio of 1:20 (400 rpm for 20 min).*
2. *Clean bare gold electrode with acetone under sonication for 5 min.*
3. *Form a self-assembled monolayer of SH-ssDNA-Biotin sequences (10 µm) on a gold electrode (3 hours).*
4. *Wash the electrode with deionized water and dry using nitrogen gas.*
5. *Prepare the electrode with a 1:1 ratio of MB-STV-AuNP (1 hour).*
6. *Pre-assemble 30 nm Cpf1 with 36 nm of crRNA at 37 °C for 10 min.*
7. *Immobilize the crRNA-Cpf1 complex on the electrode for 10 min.*
8. *Incubate target DENV-4 RNA and cleavage buffer [20 mM HEPES (pH 7.5), 150 mM KCl, 10 mM MgCl2, 1% glycerol, and 0.5 mM DTT] in the modified electrode.*

9. *Add exonuclease 1 with a proportional reacted condition for 1 hour (RT).*
10. *Calculate electrochemical data as a ratio of the delta current [I = (I – I0/I0)] before and after cleavage, where the difference in peak current is proportional to the target DENV RNA concentration.*

Appendix 8: Method for Cytokine ELISAs

Human Magnetic Luminex® multiplex Screening Assay (R&D Systems, Minneapolis, Minnesota, USA) based on flow cytometric analysis of magnetic antibody-coated beads specific to the analytes of interest can be used.

1. *Antibodies specific to the 21 selected cytokines are pre-coated onto color-coded magnetic micro-particles and the micro-particle cocktail is diluted 10 times prior to adding into a provided black-walled 96-well micro-plate.*
2. *Standards are diluted 3-fold while the cell culture supernatants are diluted 2-fold with calibrator diluent before adding into the wells in duplicates.*
3. *The plate is sealed and incubated at room temperature for 2 hours on a horizontal orbital micro-plate shaker set at 800 rpm.*
4. *A magnetic micro-plate housing is used to place the micro-plate during the washing step to remove unbound substances, and a biotinylated antibody cocktail specific to the analytes of interest is added to each well and incubated at room temperature for 1 hour.*
5. *After removing any unbound biotinylated antibodies, streptavidin-phycoerythrin (PE) is added to each well to bind and detect the biotinylated antibody. After a final wash, the micro-particles are re-suspended in a buffer and read using a flow-based Luminex® 200 system (Luminex Corporation, Austin, TX, USA).*
6. *Two spectrally distinct light-emitting diodes (LEDs) illuminate the beads, first LED identifies the detected analyte and the second LED determines the magnitude of the signal derived from PE, which is directly proportional to the amount of analyte bound.*

7. *The readings obtained are then analyzed by xPONENT 3.1 Rev 2 for Luminex 100/200 software (Luminex Corporation). The level of these cytokines is calculated in pg/ml determined from the standard curve constructed from the seven standards included during the assay.*
8. *The coefficient of variance of the replicates is calculated to certify the precision and repeatability of the assays.*

Appendix 9: Method for Metabolites Detection

1. Extraction of metabolites by protein precipitation:
   - *1 part of serum in 3 parts of methanol and ethanol mixed in a ratio of 1:1 at –20 °C*
   - *Vortex and incubate at –20 °C for 5 min*
   - *Centrifuge at 16,000 × g at 4 °C for 10 min*
   - *Remove pellet*
   - *Supernatant filter through a 0.22 µm nylon filter*
2. *10 µl of the extracted serum sample was subjected to a reversed-phase column (Agilent 959764-902, Eclipse Plus C18, 2.1 × 100 mm, 1.8 µm, 600 Bar) (Agilent Technologies, Santa Clara, CA, USA) thermo-stated at 40 °C using 6550 iFunnel Q-TOF LC/MS system (Agilent Technologies).*
3. *Compounds present in the sample are separated according to their polarity through gradient elution with solvent A (water with 0.1% formic acid) and solvent B (acetonitrile with 0.1% formic acid) at a flow rate of 0.6 ml/min.*
4. *Start the gradient from 25% B to 95% B in 35 min, return to 25% B in 1 min, and re-equilibrate at 25% B for 9 min (Ciborowski et al., 2012).*

   At the same time, samples are subjected to electrospray ionization (ESI) in positive ion mode that converts the compounds into positive ions.

   The mass analyzer then resolves these positive ions and fragment ions in a quadrupole time-of-flight tube where the ions were accelerated by an electric field consisting of four equal monopoles with alternating polarity.
5. *The mass-to-charge ratio (m/z) of an ion is then measured by a detector operated in a full scan mode from 50 to 1000 m/z based on the time used to travel across the tube.*

The molecular masses of the ions are determined and subsequently, the metabolites were identified through MS/MS spectra (Pitt, 2009; Zhou, Xiao, Tuli, & Ressom, 2012).

6. *Raw data files obtained from QTOF-LCMS were extracted using an untargeted batch-processing feature extraction software, MassHunter Profinder (Agilent) to generate processed .cef file to be input into a chemometrics software, Mass Profiler Professional (Agilent) for data analysis (Jenkins, Fischer, & Sana, 2013). Unsupervised principal component analysis (PCA) is performed for quality control of the samples based on the variability in the data and their possible correlation.*
7. *The included data set was then reordered and grouped into HC, DWOWS, DWWS, and SD. Data was prepared by filtering the frequency and abundance, aligning the parameters based on tolerances established by retention time and mass, and normalizing to reduce the variability caused by sample preparation and instrument response followed by base-lining the statistical abundance across all the samples for further data analysis.*
8. *Fold change of the metabolic expression in dengue groups (DWOWS, DWWS, and SD) against HC was calculated and statistical analysis was performed using multivariate Welch's one-way ANOVA with Benjamini Hochberg false discovery rate for multiple group comparison. Selected metabolites were highlighted by calculating the prevalence of expression among the categories along with the abundance of ion detection representing the level of expression in each individual.*
9. *Visualizations such as Venn diagrams, volcano plots, heat maps, histograms, and boxplots have been constructed to present the findings of the experiment. Pathway enrichment analysis was also completed based on BioCyc database collection (Caspi et al., 2016) to identify metabolic pathways involved in the progression of dengue virus infection.*

Index